Tom Song-Thrush
and
the Big Hush

By Barry J Tindall

waterswhelmingbooks

Tom Song-Thrush and the Big Hush
© Barry J Tindall
Published by Waters Whelming Books
First published in 2020
ISBN - 978-1-84-499085-6

© Text copyright Barry J Tindall
Design by Camilla Lovell
Illustrations by Nataliia Tymoshenko
Printed in Great Britain by Halcyon Print Management Ltd

Web site: www.tomsong-thrushandthebighush.co.uk

Email: tomsemail@tomsong-thrushandthebighush.co.uk

Tom Song-Thrush
and
the Big Hush

By Barry J Tindall

Foreword

How I got to write this book.

Well ....

As a child growing up in the countryside I spent most of my time, when not at school, wandering the local hills and fields. I developed a passion for birds and was never happier than being outside watching and listening to them.

As I grew older I became distracted and there wasn't time in my busy life for my favourite thing, until .... A lot of years later!

I moved back to the house I grew up in and immediately set about reconnecting with my younger days and my birds.

It was springtime and they were in full voice. It was fabulous and I became entranced with the multitude of chirps, twitters, coos, tweets, warbles and whistles. It was like they were talking to me.

However, I was also upset that many of the species I knew and loved as a child had disappeared or at least dwindled in numbers.

Species like Lapwing, House Martin, Pied Wagtail, Grey Partridge, Spotted Flycatcher, Cirl Bunting, Song Thrush, Starling, Dunnock and others ... all had gone completely or were seriously avoiding me. That's far too many!

We have contributed to this decline by interfering with their habitat and changing our farming methods.

Anyway, the more I listened to birdsong the more I became convinced they were trying to tell me something. There was one particular Song Thrush who used to sing his heart out morning, noon and night. It is a beautiful song and in my mind he was saying ..

  "We need help, please listen!"

He was clearly going to be the "spokesbird". I called him Tom and wrote this book about how he and his friends tackle the conservation problems they are facing.

I hope you enjoy it!

Barry the (reborn) Birder

Acknowledgements
"I would like to thank the birds for sharing their concerns and inspiring me. Without them this book would have not been possible"

To tweet or not to tweet

Councillor Dunwell spoke.

"So," he said slowly, choosing his words very carefully, *"I have listened to various stories, read many emails from the local community and watched your curious behaviour."*

There were hundreds of birds assembled in front of the village hall but he was only looking at me, Tom Song-Thrush.

The councillor carried on. *"Strange as it may seem, I am forming the unprecedented opinion that you birds may understand what humans say. With this in mind, and completely relevant to this meeting, I have a question for you."*

I continued looking at him, not daring to look away. All my compatriot birds did the same. This was the climax of all our work and sacrifice. Had it been worthwhile, had we made a difference? Our hearts were in our beaks.

He carried on, and looked straight at me. *"I would like to ask you this, Mr Thrush ... or is it Mr Song-Thrush?"*

I was bursting, I wanted to answer, I wanted to say *"It's Tom Song-Thrush actually"*, as I thought two surnames sounded more important and now was the time to sound important!

But I couldn't. Councillor Dunwell was right, we can understand Humanish language, we can't speak it!

He continued: *"My question is:"*

We were all on tenterhooks. When he next spoke, we knew what to do.

Eighteen months earlier

Chapter one
April - Nestling Tom

Hi,
My name is Tom Song-Thrush.

I may not be the handsomest or brightest-coloured bird in the world but boy can I sing. Early morning or late evening, the bird at the top of the tree making all the noise is probably me.

I was born and raised in a place called Elmdown Wood. It has always been my home and is surrounded by fields, hills, streams, a big river, more woods and a small village called Little Snuffling.

Little Snuffling is inhabited by the unfeathered, which is our name for humans, and because of some bizarre twist of fate we can understand their Humanish language, but can't speak it.

If they knew we understood them life would be very different!

Elmdown Wood and the surrounding countryside are beautiful and I am lucky to live here. But I am troubled …

I have a pretty good memory. My first one is of me as a nestling Song-Thrush.

I'm in my nest with beak permanently wide open being fed insects and worms by mum (Tilly) and Dad (Tom Senior). It had to stay permanently open as there were four mouths to feed including mine and closing it for a moment might mean missing a meal!

I shared the nest with my brother Tim and sisters Tippi and Tessa. There wasn't much room and the competition for food was great. Mum and Dad were super busy keeping us all fed and we were growing fast. I had a particular liking for worms but took what I could get as my siblings didn't seem that fussy and ate anything in a parental beak.

After a week or so of frantic feeding, I remember looking over the side of the nest and thinking **What's next?** I want to get out of here as there's no room any more and I have a strange urge to spread my wings. At the time I had no idea how true that was!

But I was six feet up in the air.

How was I going to do that?

I posed this question to Mum and Dad, who explained to me that the strange stubbly things growing off the side of my body were wings and feathers and that one day they would allow me to fly.

Fly?! I thought.

Fly?! Imagine being told that … **woohoooo!!**

What they didn't tell me was that being able to fly came after the evacuation of the nest, so an uncomfortable first landing was inevitable.

After another week of frantic and competitive feeding it was time to leave the overcrowded nursery. Basically, it's a leap of faith as my wings weren't ready for any kind of flying. So I jumped, flapped like crazy and plummeted to the ground with a bump. I shook myself down, checked for damages. No harm done though, just a bit shaken.

The next few days I spent walking around with my beak wide open, eating a lot, courtesy of Mum and Dad and waiting for my new feathers to grow enough so I could get airborne.

It was during this short time that the seeds were planted in my brain that were to have a profound effect on my life. Dad was a great storyteller and once a day he used to gather us round for one of his yarns. I always looked forward to them more than my siblings. His stories were always amusing and entertaining and consisted of tales about his younger days including meeting Mum, adventures with friends, trouble with the local unsavouries (bird predators) and what a perceptive fellow my granddad was.

But it was the old stories he told that had been handed down through the thrush generations that I found most interesting and they gripped me in a powerful way. To me, the time-honoured tales made both fascinating and pretty grim listening. Basically, the gist of the stories were that we Song-Thrushes were slowly dying out because of changes made by the unfeathered to the local habitat. Dad stories told of days when the fields were smaller and there were a lot more hedgerows. The unfeathered weren't so obsessed with tidiness. Field edges and grass verges with wildflowers were left to grow naturally. The farming unfeathered were happy with smaller fields and smaller tractors. This was great for us birds as the environment encouraged wildlife to thrive and provided us with lots of food and more places to nest!

I loved my Dad, and generally he is an upbeat bird who always encourages us, telling us to follow our dreams, keep a smile on our beaks and always try to be happy.

Having said this, Dad always seemed to make storytelling something of a warning about the future and used phrases like: **"We thrushes are diminishing in numbers you know son"**, or **"There's not as much food now as there used to be"**, or **"The unfeathered keep taking our homes for their tractors"**, or **"If they keep destroying our habitat us Song-Thrushes will be no more."**

This is the troubling bit I mentioned earlier.

Chapter two
April - Flying fun

So I learnt to fly.

And it is as amazing as it sounds. One day you're bimbling around on the ground with short stubby underdeveloped side attachments and the next day you flap them and you're up in the air! It takes a bit of getting used to, with early manoeuvrability tricky and the landings somewhat painful, but with practice I became a proper, flying bird. There is nothing like it! Let me try and describe this.

Imagine for a moment ...

You are sitting in a hedge, snacking on the occasional insect that strays too close and looking for something to do. You spy a tall tree across the field and think, I bet the view would be good from up there! So without another thought, you just go. Boom, that's it, flap, flap, through the air, no obstacles, to the new place. It's so great!

And as if flying wasn't cool enough: when I was still a nestling I used to listen to Dad singing in a nearby tree. I was amazed at what a fantastic voice he had. So loud and clear with every song not quite the same as the last song.

What I didn't know was that all of us Song-Thrushes, especially the boys, can sing like my dad! The extra cool bit is, and sorry to sound vain but I've been told, even for a Song-Thrush, I have a naturally good voice.

I'm a lucky bird.

Chapter three
May - Rescuing Roscoe

Growing up in the countryside, life was pretty good and mainly trouble free. In my younger days I had lots of fun with my mates. Racing across the fields next to Elmdown Wood and hurdling the hedgerows was a favourite pastime and pretty competitive.

Another fun thing we used to do was to see how quickly we could fly over a field we called Pongside. Pongside was inhabited by lots of unfeathered-owned pigs, who pong. The challenge was to do the whole flight in one breath as they did hum a bit.

Bill Blackbird is my best mate. He was a cousin of mine, slightly bigger, all black with a bright yellow beak. He was brave as a lion, very loyal and an excellent hunter of worms due to his supersonic ears. He was also a fine crooner. We used to compete with each other in the hedgerow steeplechase and the Pongside dash. Bill usually had the edge in the steeplechase when out-and-out speed was required. His wings were slightly longer than mine so he could fly a bit faster. However, I used to be able to do the Pongside dash in one breath quite easily. Singing loudly for hours gave me good, strong lungs. To even things up Bill was fearless, and I had fears aplenty.

It was following a victorious Pongside dash that I met Roscoe. As you will see later in the story, this proved to be quite fortuitous.

I had alighted on the branch of a big pine tree on the far edge of Pongside. I turned to see Bill gasping for air as he was trying to complete the fly-over in one breath, when I heard an unusual noise. It was a "cawing", grunting sound which I thought was similar to a raven but not as deep. On investigation I found the source of the noise and I was right, it was a fledgling raven and he was in trouble.

The poor bird had just left the nest at the top of the tree I was in and made a very amateurish attempt at flying. He had flapped a bit and mostly fallen to the ground and ended up with his left leg impaled on a barbed-wire fence. Farmer Cobbled had put the fence in, to keep his ponging pigs from wandering off.

The young raven was clearly in pain and not able to speak to me coherently. I did manage to ascertain that his dad was Rafe Raven and that Rafe and the fledgling's mum had flown over to the White Clumpys to look for food.

I was tempted to fly away and leave the fledgling raven for his parents to deal with when they returned, as Ravens and Song-Thrushes aren't mates at the best of times. They'll steal our eggs and nestlings given half a chance. But I couldn't. The bird was young, suffering and frightened and I felt sorry for him. Raven/Song-Thrush incompatibilities went out of the window and I wanted to help.

I assessed the situation and decided I would try to take the weight off the impaled leg by lifting him onto a log that was beside the fence.

This wasn't as easy as it sounded. Even though the Raven wasn't fully grown, he was still a big and heavy bird. He was very nearly upside down and almost dangling from his injury. He must have been in a lot of pain. I positioned myself under his body and pushed upwards toward the log as hard as I could.

The bird was distressed and making an awful noise but luckily I managed to shift his weight. As I lifted, his body twisted and the wounded leg came free. This was very fortunate as he could have fallen lower and impaled himself further.

With a lot of huffing and puffing I managed to get him onto the log. He was bleeding badly but not sounding as distressed as before. He still looked frightened so I tried to reassure him that the worst was over and that I would fetch his parents forthwith. In the meantime, I told him to hold the wound together with his beak to stop the bleeding. I would fly over to the White Clumpys and find his dad, Rafe, and his mum whose name I didn't know.

Before following my instruction to stem the blood flow he managed to say **"Thank you for helping me Mr Song-Thrush, my name is Roscoe."**

I was relieved he was not in so much pain and felt happy that I was able to help, even though the beneficiary was someone who would happily destroy my family in a heartbeat. I didn't think about that, I had done what I thought was the right thing. You never know, it might come back to me.

I flew as fast as I could over to the White Clumpys and found Rafe. He was grateful and said he owed me one. That's the main reason I chose Rafe for security. The second reason is more obvious. He's big and strong and looks like a bouncer.

Chapter four
June / September - Summer fun with friends

Another game we had was to fly closely backwards and forwards past Jennifer Wrennifer as she sat hidden in a bush. This annoyed her, like most things.

Jennifer was another good friend. She was a tiny bird with a pointy tail and a very loud, opinionated voice. Jennifer was a bit bossy and had little patience. She even scared the brave Bill sometimes when she was particularly shouty. Having said this, she would do anything to help her friends and sometimes strangers if she thought they needed it.

But that's another story.

We viewed provoking Jennifer as harmless, mischievous fun. The fun bit was waiting for her to explode into a tirade of abuse that, to your ears, might sound like a loud and melodic song. She was small but she packed a big punch so we knew when to beat a hasty retreat when the "song" started and off we'd fly.

My sisters Tippi and Tessa never got involved in our games. They both seemed intent on growing up quickly and meeting a suitable Song-Thrush boy as soon as possible. I was shocked at how focused they both were in acquiring a partner, as it didn't take them very long!

Now they have moved away to be with their new beaus, so I don't see them much any more. However, with the wind in the right direction I can sometimes hear Humph (Tippi's husband) singing loudly. Humph has a good voice, we must perform a duet at some point. By the joyous sound of his singing, all is good in his world. This is a sign that my sister must be happy, I like this. Tessa lives further away with Thaddeus Thrush, a big and very serious bird who doesn't mix much in our family. He's also a pretty poor singer.

Myself, Bill and Jennifer had excellent singing voices and for fun we used to come together in a makeshift choir. Another member of the choir, which we had imaginatively named the "Elmdown Croon" was Rosie Robin with her laid-back flutey tones. My friend Rosie is a small but pretty bird with a lovely red breast. If I'm honest, I quite liked her in the romantic sense, but she wasn't a Song-Thrush so we remained friends.

Rosie was not a shy bird and was very inquisitive. Her inquisitiveness has in the past got her into some scrapes but I won't go into those now. She had a nice voice and liked to sing all year round, even during the night!

And our final, regular member of the choir was another girl friend of mine, Darcy Dunnock. Darcy was a very smart bird and always full of good ideas as you will see. She was referred to by the unfeathered as a "little brown job", or a hedge sparrow. She liked neither, and used to get annoyed when she heard either description. Darcy could sing beautifully but had a habit of stopping her song too soon. I have asked her about this and she says it's just her way. "Leave them wanting more," she says, whatever that means.

Sometimes we were joined by Bertie Blackcap, Will Whitethroat and Gobi and Gertie Goldfinch. These fellows are also excellent singers. When we got together and sang, Elmdown Wood was a cacophony of sound, especially early in the morning. The Warblers, Bertie and Will were very happy tweeting and chirping away. Bertie was probably a little more in tune than Will, who did sound a bit scratchy at times, but combined they were a tour de force.

Then there was Jennifer, melodically shouting more than singing in her loud voice, and Darcy firing up her fine song before ending it abruptly, just when things were getting good.

Gobi and Gertie provided excellent background vocals and then there was me at the top of the highest tree, performing my improvised, tuneful, sometimes repetitive and very loud song for hours on end.

We all had our unique and appealing voices and together it sounded pretty spectacular.

Because we lived near the village of Little Snuffling and encountered the unfeathered on a regular basis, it was obvious to us that they liked our singing. When out walking, sometimes with a dog and sometimes without, they would stop, listen and invariably smile whilst commenting to each other how lovely the birds sounded today. We know this because we can understand what they say! All the birds in the Elmdown Croon agreed that It was nice that they liked our singing.

During the summer I had been honing my singing skills and spent most early morning and late evenings singing loudly from a prominent position to anyone who would listen. As I mentioned earlier, the unfeathered liked our singing. I was happy knowing this, as I enjoyed performing for them.

With this in mind, and, I like to think, a smattering of selfless generosity, I decided to find a nice tree near the village so the unfeathered could hear me easier.
In reality, like any good performer, I guess I'm pretty vain and enjoy the attention.

Anyway, I found a cracking old oak tree with a couple of pointy up branches at the top which served as an outstanding perch. From there I'd serenade Little Snuffling on a daily and nightly basis. Sure enough many of them stopped, listened, pointed, smiled, laughed and sometimes applauded as I offered my full repertoire of tweets, hoots, chirps, whistles and croons on repeat. My buddies loyal Bill, smart Darcy, inquisitive Rosie and bossy Jennifer often joined me and sang in trees and bushes nearby. They were also welcomed by the unfeathered.

We were a popular show.

Chapter five

October / November – The Winter Berry Bash

Summer was coming to an end, the days were getting shorter and the temperatures were getting cooler. I was now a master pilot with many flying hours under my belt and a good knowledge of the surrounding area including a small portion of the White Clumpys which are a couple of miles away. The White Clumpys is our name for a range of hills that we can see in the distance. They aren't particularly white but they are clumpy. Clumpy means an area covered in trees. The Clumpys hide lots of small unfeathered villages in their valleys. Bill and I fly over there sometimes when we are feeling bold. We are a bit wary as there are an awful lot of unsavouries in the form of red kites and buzzards. Our local red kite family is headed by Ron who we can't trust and have to keep a careful eye on. The same goes for Bram Buzzard who also lives in Elmdown Wood. Watch your nestlings or he'll have them from under your nose.

The White Clumpys are a favourite habitat of the aforementioned Ron Red-Kite and Bram Buzzards and their families so caution is required. Having said that I believe the hills could offer some great adventures.

Around October Mum and Dad told us kids to expect a visit from our thrush cousins, Rodney Redwing and Frank Fieldfare. They are both thrushes but from different families and they always come to Elmdown Wood in autumn. Rodney and Frank live in a land far away, over the White Clumpys and across a sea but it is very cold there in winter so they come to visit us. Not just to avoid freezing but also to feast on our plentiful berries! Mum and Dad always gave them a warm welcome, organising the famous **"Winter Berry Bash"**. This is an annual winter shindig that's not only a celebration of returning old friends but an appreciation of the winter harvest from the fruit and berry laden trees. I say "Thrush" Berry Bash when actually Mum and Dad invited all our bird friends from Elmdown Wood and beyond and a good time was always had by all.

Winter soon followed this time of plenty. The harvest didn't last long. Not only did all the residents of Elmdown Wood and the surrounding hedges and fields need to eat but we had Rodney and Frank and all their mates here as well.

Added to that the unfeathered picked a lot for themselves. The berries and fruits soon disappeared from the trees. In winter food was scarce, it was cold and we hunkered down.

It was November time and I was now technically grown up at seven months old. I looked forward to next spring which seemed an age away. Something told me it was going to be special.

Call it spider bird sense.

Chapter six
February -
Laurent Lapwing

It was a cold day in February and I was hopping around the edge of Elmdown Wood looking for a nice beetle or worm to munch.

I had arranged to meet Bill and demonstrate to him my skilful technique of smashing a snail shell on a stone when Dad flew over to me with an urgent look on his face.

"Tom, stop whatever you're doing now!" he said in an abrupt manner. Dad's not normally a rude bird so I thought it must be important. *"Quick, come with me son! I want you to meet someone."*

Bill had just turned up in anticipation of learning something new and possibly useful. Blackbirds are not natural snail destroyers, much preferring a wiggly worm for a meal but he thought he would go along with my desire to teach him, as we were mates. I looked at him. *"Sorry mate, better go, Dad's got a bee in his feathers about something."*

"Sure, no problem, we can continue my 'education' later," Bill replied in a sarcastic voice.

"Right!" said Dad, and flew off towards one of Farmer Cobbled's fields. He landed in the long grass at the edge of the field with me right behind him, and gestured with his wing out into the field. *"See that handsome bird over there, with the sticky-uppy bit on his head?"*

"Yes," I said.

"That is a lapwing," Dad informed me.

"Never seen one before," I said.

"Exactly!" said Dad. *"Go introduce yourself and ask him why."*

To be honest, I was already interested in finding out about this somewhat glamorous newcomer as he was a fine-looking bird. I flew over immediately. *"Hi, I'm Tom, Tom Song-Thrush,"* I said. *"How are you?"*

"Oh hello!" said the glamorous bird with sticky-up head feathers. *"My name is Laurent, Laurent Lapwing."*

I continued. *"Yes, my dad told me you were a Lapwing. I've not seen you around here before, he said I should ask you why that is?"*

"Did he now?" said Laurent. "Actually, I'm just passing through and needed a break. I've been flying for hours and I was getting a little peckish and thirsty."

"Oh!" I said. *"If you want some water there is Farmer Cobbled's cattle trough over there. I'm not sure what you eat so I can't advise in that area I'm afraid."*

"What do I eat?" said Laurent. *"Nice juicy earthworms or leatherjackets are my favourites. I'll have a hunt around and see what I can find, thanks."*

"So why have I not seen you around here before?" I asked again. *"This is a great place to live."*

Laurent had a sad look in his eyes. *"It might be for you but not much good for me and my kind. I know this place, my parents told me about Elmdown Wood and the fields near the White Clumpys."* He was getting quite talkative now. *"They told me that my great grandparents and Lapwing generations before that used to live here in big numbers."*

"That's interesting, what happened?" I said.

"Well, it's a food and habitat, or lack of, thing," said Laurent.

"Really?" I replied, becoming more engaged.

Laurent continued. *"To cut a long story short, the unfeathered farmers around here changed the way they grow stuff. They drained the land, changed the crop planting times and took away the habitat we like. Slowly but surely my forefathers realised it was time to move to a more suitable area."* He added in a half-joking tone: *"So Elmdown Wood probably isn't the ideal place for me to stop for refreshments, but it is pretty around here. Hopefully I'll find a few worms to keep me going, otherwise my onward journey might be quite difficult!"*

I was starting to warm to Laurent, and what he was saying was interesting. He carried on. *"There were lots of us Lapwings back then but not so many now and we've had to leave these parts for the reasons I said."*

"Where did you go?" I asked.

Laurent replied. *"My grandfather took our family north. He found a place where the soil is wetter and the fields are smaller and not full of corn. Much more food there."* He then looked into the distance. *"I do worry the farming unfeathered will change the habitat in the future and we'll have to move again."*

It dawned on me why Dad had wanted me to talk to Laurent. This was a real example of Dad's stories coming to life and it was powerful stuff. Lapwings had already suffered what Dad and his ancestors were worried was happening to Song-Thrushes.

"That's terrible," I said.

"Yes, not good," said Laurent. *"Anyway, it is what it is. My family has struggled in the past and have had to go exploring around the countryside to find places where we can nest safely in springtime and there are wetlands for food."* A perceptive and thoughtful Laurent then posed a question: *"Are your family struggling as well?"*

"Is it obvious?" I asked, with a grin on my beak. The grin was to show friendliness, it didn't portray how I actually felt. *"Yes we are,"* I admitted. *"I understand now why Dad wanted me to talk to you. He's trying to make me understand that we have a problem."* I was sad that Laurent didn't live around here anymore, sad for his family and sad as he seemed a good bird bloke. *"I'm sorry to hear what happened to your family,"* I said. Dad's plan had

worked, and a feeling of determination as well as a sense of purpose entered me. *"One day I am going to do something about this habitat destruction and stop the unfeathered ruining our bird environments,"* I said, with as much conviction and assertiveness as I could muster.

"Hmmmmm ... good luck with that!" said Laurent, rather dismissively. Then he looked at me and noticed I hadn't taken that comment well. *"Hey, sorry Tom,"* he said, *"that sounded a bit harsh. I had no right to say that and I apologise. You go for it Tom, do what you can to help your Song-Thrush family and all other birds."* As he continued I sensed he was thinking back to when Lapwings were declining in our area. *"We never had a plan to tackle the unfeathered farming practices, maybe you have? Lapwings may be handsome birds but creative thinking and forward planning skills are not our strength"*.

Laurent prepared to leave. *"Good luck Tom Song-Thrush!"* he said. *"Must go now, I need to find a nice, damp, ploughed field and some more Lapwings where I can settle for a bit."* Then, *"Nice to meet you!"* he said in a positive, upbeat manner as if to counter his previous negative comment. *"And good luck with whatever you decide to do for your family! I'll pop by next time on my way through from my winter home in France."*

With that he flew away leaving me pondering what he had said.

Dad appeared. *"What did he say?"* he asked.

"He said what you knew he was going to say," I replied.

Dad fixed his eyes on me with a serious look on his face. *"As you have now heard, we are not the first to be affected by local unfeathered activities. Unfortunately for Laurent and the lapwings it has already happened, their ideal habitat has gone. Ours is disappearing and the threat is real."*

I was a young, impressionable bird. Dad's old stories and my meeting with Laurent had a profound effect on me. I wondered about the future and wondered how I could help. I looked at Dad and blurted out, ***"I'm not going to let it happen here!"*** I was feeling a little emotional. ***"It's not happening here!"*** I said, getting quite animated.

Dad told me to calm down, but immediately encouraged me. ***"I like your attitude and thinking son, talk to me about any plans you have and together we will try to make a difference."***

This was reassuring, I thought. With my amazing dad's help we could achieve anything. He looked at me with quiet pride. I flew back to Elmdown Wood but couldn't stop thinking about our Song-Thrush plight. If, in the future I have kids and continue the traditional storytelling what is going to happen? If nothing is done and they pass on the stories to their kids, surely Dad will be right? One day there'll be none of us left.

I tried to console myself: hopefully, the good hearted unfeathered (of which there are a lot), will help us by protecting our woods, hedges, fields and grass verges.

Not just for us Song-Thrushes but for all other threatened birds so we can carry on singing from tall trees and hedgerows and cheering them up.

I knew somewhere deep inside me that one day I would do something to help stop our Song-Thrush decline and in so doing help other birds as well.

Chapter seven
early March - Bella love

Spring was approaching. It was my first one as an adult bird. Soon after my leap of faith from the nest, Mum and Dad explained to me and my brother and sisters what happens to birds in springtime: our feelings change and we find ourselves doing new and unfamiliar things in the hope that we might meet a nice bird to settle down with and start a family. It's all natural, we don't have to think about it. But to me it sounded complicated. I wasn't ready to grow up.

Or so I thought.

It's often said amongst grown-up birds that love conquers all. This made no sense to me until March 5th this year.

As I've mentioned before, singing is important to me. Every day at various times I will find a nice, comfortable and prominent branch in a tree and open my pipes, sometimes near Little Snuffling and sometimes in Elmdown Wood and sometimes in between. I am blessed with a beautiful voice and the songs just come. Not like learning to fly, where I had to practise. When I sing I just open my beak, blow a little and all the tunes are there. Some high, some low, some fast, some slow, some I repeat and some only ever happen once. I have no control over them! It is a gift and it gives me a lot of pleasure. I am grateful to whoever decided Tom Song-Thrush shall be a special singer.

On this particular day the weather was fine and I was in a good mood. Early March is the time when us male birds increase our singing in the hope we may attract a lady bird. I was still young and not too bothered about settling down, **I just liked singing!**

Anyway, I was in Elmdown Wood, on one of my favourite perches, quite high up in a sycamore tree with a view across the fields to the unfeathered village. I was randomly whistling and chirping and tweeting for all I was worth. The sun shone down and it warmed me and the music was coming thick and fast. I was in a good place but little did I know it was about to become a better place! At the time, I was oblivious to the fact that I had an audience. Below me, on a lower branch to the left was a young lady Song-Thrush. She was staring at me.

It was a beautiful day. There was little wind, bright sunshine, a clear sky and my voice travelled a long way. It pleased me to think that I could be heard from a long way off. As I've mentioned before, what comes out of my beak is not rehearsed and no two tunes are exactly the same. Sometimes I will finish a singing session and think it wasn't very good, but I don't worry. I know the next one will be different and, I always hope, better.

I was on fire today though. After about twenty minutes of spontaneous crooning, I decided it was time for a break as I was getting thirsty. As I flew down to the small pond in the middle of Elmdown Wood, I was thinking, I can't wait to get back up on my perch to continue my performance and hope it was as good as the last.

As I was dipping my beak in the water a voice said *"I like your singing."*

I jumped and spurted out the water. Another bird had flown down and landed next to me. I looked to my left. A young lady thrush looked into my eyes and laughed. I was taken aback and felt a little unnerved. She was a fine-looking bird and I can be a bit shy. ***"Sorry about that!" I said. "You made me jump, I don't normally spit water out when drinking."***

"I like your singing," she said again, eyes fixed on me. ***"My name is Bella."***

Bella, that's a nice name, I thought. ***"Why, thank you Bella,"*** I said. ***"My name is Tom and it's the one thing I can do well."***

"I don't believe it's the only thing," she said, eyes growing wider as she spoke.

I thought for a moment. I thought about three things:

A – Was I good at anything else?

B – She is very pretty, I'm feeling intimidated.

C – Should I tell her what else I think I'm good at?

I went with C.

"Now you come to mention it, I'm pretty good at opening snail shells by banging them on stones," I said romantically.

She laughed. I'm not sure why she laughed. Maybe the romantic notion of broken snails or maybe she just liked my face and was happy to have met me? I hoped that was the case. It was a sweet laugh on a pretty face. Today was getting better.

Bella went on, *"I live across the fields and over the big farm track in a small copse. A friend told me there was a handsome young Song-Thrush in Elmdown Wood who had an amazing voice. Is that you?"*

By now I was bird blushing. *"Yep that's me, and your friend was right!"* I said in an attempt at humour.

It worked: Bella laughed.

"You flew over here to listen to me sing?" I said, in a surprised fashion.

"Might have done," replied Bella in a coy way.

"I hope it wasn't a wasted journey!" I said.

"No, you passed the first test," she replied.

I was thinking, First test? What's that about?

I decided not to pursue this train of thought as it was distracting, so just went back to the moment. *"Great!"* I said, *"Do you like our wood?"*

"From what I've seen, yes it's very nice," Bella replied.

I actually had a plan for today. After finishing singing I was going to fly over to Little Snuffling and sit on the unfeathered's church roof and listen to their music and singing. I was now busy thinking how I would like to include my new, pretty friend in this plan. So I thought for a moment then blurted out. *"If you're not busy would you like to fly over to the unfeathered village and sit on their church roof with me?"*

Bella, with a quizzical look, said *"Well, I'm not busy so yes, why not. Why are we going to sit on the church roof?"*

I replied *"good question!"*

Bella, her head to one side was listening intently. *"I go there regularly, I can hear the unfeathered sing and play their musical instruments below and I like to listen. It gives me inspiration."* Her head seemed to have moved further to one side. Either she was really interested or falling asleep.

I continued. *"The next time I sing, notes and melodies from the church music appear in my tunes. I don't know how but they do. I think I'm the only Song-Thrush who does this. It gives me the edge I reckon."*

Bella was impressed. ***"Well, you're certainly the best singer I've heard, except for Nige Nightingale. But he's not important."***

I tried not to be offended and wasn't. I know these nightingale chaps are pretty special.

We flew to the church and had a wonderful day listening to the unfeathered music, chatting about our families and favourite things to do. I hadn't felt like this before about another bird. Of course Bill was a good and handsome friend with an excellent yellow beak but this was different. I liked the feeling. We'd only met a couple of hours ago and everything was great! I wondered what she was thinking?

As evening approached Bella said it was time for her to go home. I understood but was a little sad as I was having such a good time. ***"Can I fly you home?"*** I offered.

"No thanks," Bella replied.

I wasn't finished: before she flew off I put myself completely out on a limb and said, ***"Can I see you again?"***

Bella turned and looked at me, in the same way she did when we first met. ***"Of course"*** she said, in a comforting way that implied there could be many more meetings.

Off she flew back across the field and over the road to the small copse. I had a feeling this was a journey I would be making regularly from now on. I was feeling good.

What a day, March 5th.

Life would never be the same.

Chapter eight
mid March -
Growing up

Things moved fast. Bella and I met again and again and then again. Pretty soon it was clear to both of us that we wanted to live together and most probably build a nest and start a family. We announced this to Mum and Dad and then to Bella's parents Thierry (Bella's dad is from France) and Thelma Thrush. Mum and Dad were thrilled and probably a bit relieved and Bella's parents the same.

Now, being a Song-Thrush who likes to party, this was an excellent excuse to have one. Fortunately Bella likes to let her head feathers down as well so we organised a celebratory bird bash.

All Elmdown Wood birds were invited, with the exception of what we call the "unsavouries", birds including Ron Red-Kite, Cain Crow, Meg Magpie and Sam Sparrowhawk. Birds who would do us songbirds harm. Party poopers and worse the lot of them.

With that in mind I needed some security and thought of Rafe Raven. Rafe could himself be a bit of a party pooper but he did owe me a favour from when I helped young Roscoe. Despite his threatening look and questionable reputation, Rafe was an honourable bird and agreed to help. He and a couple of heavy raven mates made sure there was no problem from the aforementioned ne'er do wells.

The party was outstanding. We drank lots of elderflower wine which was kindly donated by our friend Charles Chaffinch. He had a tiny vineyard on the south facing side of Elmdown Wood, and we danced until the sun came up over the White Clumpys.

The music was provided by **"Will's Warbler Collective"** – a band including Chip Chiffchaff, the aforementioned Will and a couple of other warbler friends. Great dance music for us birds. This was Will's main band, singing in the Elmdown Croon was a part-time thing.

Bill, assuming he was going to be best man, made a speech, even though it wasn't the wedding. I didn't mind, he was definitely my best mate and a shoo-in for the position.

Both Bella and I had headaches in the morning due to over-indulging in Charles's excellent tipple, but agreed the party was brilliant and that it was a fine way to celebrate our future plans together.

I started to feel a bit grown up now. I'd fallen in love and had committed myself (almost) to one bird.

I'd better start thinking grown up … I was also thinking, How do I do that?

Chapter nine
late March -
Losing Dad

This bit is hard.

It was late March and in the space of a few weeks I went from meeting Bella and feeling on top of the world to experiencing a horrible darkness that was truly the worst thing that happened to me.

It was a normal day and Mum and Dad were going about their daily business, flying around the Elmdown Wood, seeking out insects and beetles in the undergrowth. Every so often Dad would perch high up on a branch and exercise his vocal chords.

I could usually hear him from wherever I was. It gave me a feeling of security when I heard Dad singing. After his second session of the morning he announced to Mum that he was going to fly over towards the White Clumpys and explore a few hedges over there. He had heard there were a couple of crab-apple trees growing which, if true, would be a good source of food for the family during the colder autumn and winter months.

Mum said she was going to stay local and would see him later.

The hedges in question were on the other side of Elmdown Wood, across two fields and over a road. A place where the unfeathered drive their cars.

Dad arrived and started looking for the potential new food source.

Soon he had found the crab-apple trees in question. He was chuffed as they were blossoming well which meant a good crop of the small apples should appear later in the year when they needed them.

Dad was about to fly back across the fields towards home when he spied his friend Jack Jackdaw the friendly crow. He was flying overhead when Dad called out. Jack heard him and flew down to the crab-apple tree for a chat.

"What are you up to?" said Jack.

Dad replied *"I've just found two nice crab-apple trees full of blossom. The following fruit will do me and my family well during winter. Crab-apples aren't your thing are they?"*

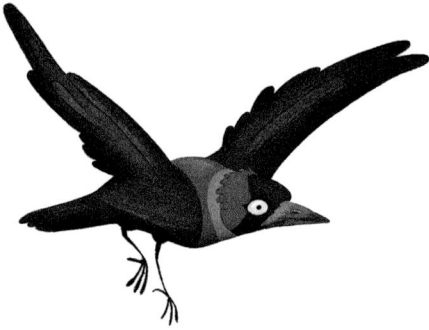

"Not for me," said Jack, *"I much prefer a bit of protein like an insect or worm. Anyway, how are you?"*

Dad replied, *"Very good thanks, life seems to be pretty good right now, everything seems to be falling into place."*

He told Jack he was looking forward to telling Mum and the family about the winter food supply and how he had grown close to me recently which made him happy. He also planned to tell other fruit-eating birds and their families about the forthcoming harvest. Bill Blackbird and Rosie Robin's parents would welcome the news.

"Nice to see you Jack," said Dad. *"I'm going to head home now, we'll catch up again soon."* And with that he flew off back across Farmer Cobbled's wheat field looking forward to imparting good news and leaving Jack to witness the terrible event that was about to unfold.

As Dad looked ahead he saw a small hedge approaching. For some reason it reminded him of when he was a young bird, before he met Mum. He used to have such fun racing with his mates across fields like this. They'd stay as low as possible as they approached the hedges then and at the last minute would rise up and over to continue across the next field to the next hedge.

Dad had shown me this game and I loved it as well.

He was in a buoyant mood. Life was good. His eldest son (me) had met a lovely bird and was settling down. He lived in a nice place with a lovely and loyal partner Tilly. As he flew towards the small hedge he was living his best life. He was looking forward to returning home with good news about winter food supplies that he could share. This made him feel good.

He was almost at the hedge now, probably flying too fast.

Dad remembered how good he was at leaving it until the last minute before rising up, clipping the top of the hedge and then back down to just above the ground, before continuing to fly across the next field.

Normally he left all competition in his wake, but today he was on his own. However, in his mind he was leading the race as a young Song-Thrush without a care in the world.

The hedge was upon him, his timing was perfect. Up he rose and flew fast between two small twigs that protruded from the top of the hawthorn hedge.

The years rolled back as he made the manoeuvre.

Catastrophically Dad hadn't realised that on the other side of the hedge was the road I mentioned earlier.

Farmer Cobbled was driving home to his farm in Little Snuffling. He had been in Bulgingtown, the nearest big unfeathered place where he had been conducting some business. Dad's timing as he rose to clear the hedge was perfect but his timing leaving the other side was fatal. He crossed the road just as Farmer Cobbled's car was at the same spot.

Dad had no chance. He was hit hard, a glancing blow from the windscreen, and somersaulted over the remaining width of the road and landed heavily on the verge. He was dead.

Jack said he saw Farmer Cobbled stop, get out of the car and go over to where Dad lay. He looked down and realised it was too late. Apparently he looked a bit shaken and gently picked Dad up and put him on the grass verge.

He stood over Dad for a couple of minutes not saying anything. Noone knows what he was thinking but as a man of the country he knew who Dad was and that Song-Thrushes were becoming rarer. He was also aware that farming practices had something to do with it.

When Jack broke the news to Mum he was in tears and Mum was beside herself with grief. Then Jack and I had a deeply saddening conversation when he told me how content Dad seemed. It broke my heart.

I can't describe my sadness, which was shared by Mum and my brother and sisters. I had lost my Dad, someone I was going to do great things with. My world had changed forever and as a family we grieved deeply.

Bella was sad as well and helped me but she didn't understand the woe as it wasn't her own dad. Having said that, I couldn't have got through the pain without Bella's support.

In the following days and weeks I thought about Dad all the time and kept coming back to the chat we'd had following my meeting with Laurent Lapwing. I couldn't forget how Dad looked at me with pride after I made my defiant rant about saving the Song-Thrush population.

Together we were going to make a difference. Not any more. It's just me.

My mission now was to make a difference for Dad.

Chapter ten
early April -
Being a dad!

My heart was heavy for the longest time but slowly I emerged from the dark place, and Bella, bless her heart, helped immensely. She was obviously sad but tried to help me stay positive. I was moving into a new, more grown-up phase in my life. Bella was busy planning nests and involved me a lot in choosing the location, the type of tree and whether it should be a high or low nest. I was happy to be involved at this exciting time, although there was always a tinge of underlying sadness now Dad had gone.

When not with Bella, I was singing a lot and trying to stay happy, as Dad would have wanted.

Best mate Bill was in a more advanced family planning situation than me. He had met his partner Buffy before I'd met Bella and had found a nest site and had built a nest. He was on strict nest maintenance and partner support duties as she brooded their five turquoise green and brown speckled eggs. This meant he wasn't available as much for hedge racing (I wasn't going to stop doing this, in honour of Dad) or worm safari excursions as he used to be.

I spent a lot of time entertaining the unfeathered with my singing from the big oak overlooking Little Snuffling. I was in good form and the melodies just flowed. My friends Darcy, Rosie and Jennifer accompanied me.

Sometimes Bill joined us but he was becoming more "henpecked" as Buffy was brooding their eggs. She insisted he stay close to her in case she needed anything. Bill being a loyal, decent and caring bird duly obliged.

From my vantage point in the old oak, I could see quite away across Little Snuffling and was aware of quite a few unfeathered relaxing in their gardens with their families. I'm sure that a lot of them were subconsciously listening to me and my friends singing.

I thought I would test this theory: I stopped singing and hid in another tree and told my friends to do the same. I kept a careful eye on the unfeathered and after a while they started to look around and over at the old oak.

I think they missed us!

Spring arrived and Bella made it clear she wanted to build a nest in the carefully thought out location we had chosen, brood our eggs and care for our nestlings. I surmised that this might be a good idea for two reasons.

A – My best mate Bill was further down the same road so the fun was over. I'm joking here, I looked forward to mine and Bella's future together with excitement.

B – More importantly, I loved her and would do anything for her.

Bella announced one Sunday morning that she was almost ready to lay eggs. This was something of a shock as we weren't officially recognised as a bird couple and we didn't have a proper nest.

Wake-up call!

We hastily arranged our ceremonial joining of which I won't bore you with the details. Suffice to say it was like the engagement party with a bit more pomp, better elderflower wine and a longer speech by Bill.

It was presided over by Osbert the Old, a tawny owl and local legend who'd been in Elmdown Wood for years. In fact all of his life. He was a good and wise bird who was generous with advice if you could be bothered to ask.

Importantly, any bird couples wishing to build nests together and lay eggs must go through a ceremonial joining officiated by Osbert.

It's the law.

Mum cried because Dad wasn't there, as did I and my brother Tim and sisters Tippi and Tessa. Some of my tears were of happiness of course as I was marrying the love of my life.

The wedding came and went and I was working hard on building a nest. Bella had found the spot she wanted, six feet above the ground in a mass of ivy on the side of a beech tree. It was a good spot offering excellent views across the woodland floor, and the ground beneath was a thick carpet of bluebells. I remembered the bump I had when I first left my nest. The bluebells would offer a softer landing for our youngsters.

I was back and forth from the field to the nest site all day with bits of straw, hair and feathers. Our new home was taking shape, Bella worked with me and between us we had soon constructed a serviceable nest.

Not a moment too soon, as Bella promptly laid in the nest four beautiful sky blue eggs, dotted with black spots.

I was blown away. In these exquisite eggs were my nestlings. I grew up a bit more that day.

Bella patiently sat on our eggs keeping them warm. Inside our nestlings were growing fast. It wouldn't be long until they were too big and would want to get out.

I supported Bella by bringing her food regularly and entertaining her with my sharp wit and funny jokes. Her response generally, in a tired voice, was **"Jokes not welcome, food welcome."** She wouldn't let me brood our eggs, I was just a food bank and general dogsbody. I must add I was happy to be this, and was looking forward to seeing our nestlings.

In between food runs and general skivvying, I kept up my early morning and late evening musical performances, mainly from the old oak by Little Snuffling. I was on a roll, the prospect of being a dad energised and inspired me. The songs were coming thick and fast and somehow I was able to incorporate more notes and sounds from my time spent listening to unfeathered church choir music.

Watching your nestling come into the world is the most special thing.

A beak appears through the shell of our egg and slowly the hole gets bigger. It's tempting to help the little chap with a well aimed peck but Bella made it clear that I shouldn't interfere. She was right, little Tim (named after my brother) emerged healthily and promptly opened his mouth wide and kept it like that for the next two weeks.

He was followed by Ben, Josie and Jill. Luckily for us all in good shape!

I was so proud and I grew up a little bit more.

mid April –
Farmer Cobbled and the council plans

As I've said before, when I do my tree-top singing it's mainly over at Little Snuffling and I'm sometimes accompanied by friends. What I didn't know was a couple of my friends, the inquisitive Rosie and the skulky Jennifer, used to explore the village further when not singing with me.

You remember I said birds can understand Humanish?

Well ...

On two separate occasions Rosie and Jennifer stumbled upon unfeathered conversations that they didn't like the sound of.

Rosie Robin was doing her thing with Arthur Ablebody, dashing in and out between his digging fork, grabbing a worm and flying back to a nearby redcurrant bush to eat.

Arthur loves this and was smiling away to himself and talking to Rosie as if she could understand. Which of course she could, but he had no idea!

It was then that his friend Mrs Rightly-Soe arrived and started chatting to Arthur.

"Have you heard?" she said.

"Have I heard what?" he replied.

"Farmer Cobbled has a plan to take out all the hedges around Little Snuffling and Elmdown Wood and mow all of the field edges to make much bigger fields."

Arthur replied, *"No, I hadn't heard about this, what brought that on?"*

"He reckons he can make more money growing crops in bigger fields," Mrs Rightly-Soe replied.

"I think this would be a mistake and such a shame," said Arthur, *"it would ruin the landscape and affect lots of birds and other wildlife."*

"I agree, it would," said Mrs Rightly-Soe. *"I don't suppose there's much we can do about it though."* She was sounding sad now.

"Let's see what happens," said Mr Ablebody, *"I think this is a bad idea as well and I will complain if I get the chance."*

"I'm with you," said Mrs Rightly-Soe, *"I will complain also."*

Rosie couldn't believe her ears. She was trying to process what she had heard. Less hedgerows, bigger fields and mowed field edges meant less wildflowers and grass. Which in turn meant less food and nesting places. This is not good for Robins!

The second conversation involved Jennifer.

She loves skulking around in bushes and ivy walls, staying hidden and watching and listening to what's going on.

Recently she was opposite Little Snuffling village hall listening to some unfeathered chatting as they waited at the bus stop.

What she heard was similar to Rosie, in that it wasn't good news.

"So that's great news!" said a local man called Johnnie Cotton. He was talking to his mate Dick Moss. Johnnie lived in a big house, drove a fancy car and was notorious for being loud and unpleasant. He was waiting for a bus to Bulgingham to collect his big and expensive automobile, which was being cleaned. Johnnie had a thing about cleanliness and tidiness. *"The council has bought new, big fancy mowers to mow all of these untidy grass verges with their wild unkempt flowers."*

"That is good news, Johnnie," Dick Moss replied. For some reason Dick held Johnnie in high esteem and always agreed with him.

Johnnie went on, *"This village is untidy, it's as if no one cares, I'm pleased we're getting things sorted out and tidying up the untidiness."*

Dick replied in a subservient way, **"I agree Johnnie, we live in a very pretty village which is being ruined by the untidiness. Short grass and sculpted, clean cut verges is what we need. We deserve it."**

Jennifer, like Rosie before, was processing this information and getting in a bit of a flap as she wasn't the coolest customer. No grass and wildflowers on the roadside verges? What about all the flies, insects and beetles I eat, what'll happen to them, they'll all disappear!

This is not good for Jennifer Wrennifers.

Both Rosie and Jennifer knew that these plans were worse for me as a Song-Thrush as we were suffering more than most. They dreaded telling me but knew they should.

At this time I was one hundred percent immersed in being a new dad to our four new nestlings – little Tim, Ben, Josie and Jill. It was hard work but my heart sang. The little monkeys were growing fast and permanently hungry.

Despite this and with the consent of my darling partner Bella, I did manage to sneak away for a bit in the morning and evenings to do some singing. It was during one of these excursions that I bumped into Rosie and Jennifer as well as Darcy, who were returning to Elmdown Wood.

They congratulated me on my new family and I assured them that one day they would have their own. They seemed to ignore this comment as they had some news that was important and had to be taken seriously.

Rosie told me of her earwigging Arthur Ablebody and Mrs Rightly-Soe's conversation about big fields, fewer hedgerows and mowed field edges.

I was shocked. If you could see a Song-Thrush's colour drain from their face, you would have seen mine do so.

Darcy also looked aghast: as a "hedge sparrow" she would miss the hedges!

Worse was to come, Jennifer then blurted out what she heard Johnnie Cotton and Dick Moss talking about: that the council had acquired new mowers to keep the grass and wildflower verges cut short.

On hearing the second piece of bad news, time sort of stood still for a bit. After a few seconds of frazzled thought, I quizzed the girls again about the eavesdropped conversations.Both were absolutely sure of what they heard.

I began wrestling with a number of thoughts:

A – Dad's stories about the declining Song-Thrush population were coming true, and

B – I'm a parent now and need to protect my new family.

The final thought jostling in my brain with the others was the vow I made to myself when Dad was killed. It was time to stand up. I had to do something not just for my family but all bird families ...

But what?

Chapter twelve
May -
The master plan

It was May time and little Tim, Ben, Josie and Jill were doing well. This stimulated my thought processes. There was no doubt in my mind that Song-Thrushes like me were reducing in numbers because of our changing habitat and environment. My mind was made up. I must do my utmost to stop further unfeathered interference and hopefully create an opportunity for us to increase in numbers.

And as Laurent explained before, it's not only us who are affected. Nige Nightingale's family has moved out. I met Nige when he was passing through recently. He told me a similar story to Laurent's. His family couldn't live here because the habitat and food was gone. Nige has the most amazing voice, probably better than mine. It had now disappeared from Elmdown Wood.

I have heard stories about Colin Cuckoo and his family who used to live around here. They've now moved on as their favourite caterpillar delicacy has disappeared due to habitat change.

Colin's voice is very popular with the unfeathered. It goes something like this: **"Cuck-oo, Cuck-oo"** followed by **"Cuck-oo, Cuck-oo"**.

As you can see, previous generations of Cuckoo have put a lot of thought into it.

Colin Cuckoo and his relatives are nasty pieces of work. I know for a fact that Darcy's family have, in the past, been traumatised by the presence of the Cuckoo clan. They are unbelievable. Mrs Cuckoo lays her egg in another unsuspecting bird's nest along with the existing eggs. The unsuspecting mother bird broods all the eggs but it is Cuckoo's egg which always hatches first. The new nestling Cuckoo then throws the other eggs out of the nest and stays in the nest on its own to be fed and watered by Mother until it can fly.

Mother is seemingly none the wiser, such is her maternal instinct. I reckon she must be a bit dim not to realise something was amiss.

To be honest the decline of Colin and his Cuckoo chums is probably a blessing in disguise. We don't miss them. The unfeathered do.

Most of my time was spent thinking about farmer Cobbled and the council's plans. It was a little disturbing to think that it was him who had killed Dad. I didn't bear a grudge as I knew it wasn't his fault but now he was planning to threaten our future. This would be his fault and would affect my close family, and Song-Thrushes in general, as well as most other birds in the area for generations to come.

I told Bella and she thought the same as me. **"You must do something,"** she said with succinct clarity and followed up with, **"and you will do something."**

It was encouraging that she believed in me but … What could I do? I repeated in my head.

I took it upon myself to call a meeting. It was to be held at the clearing in Elmdown Wood and all birds were invited and I hoped, would attend. As a Song-Thrush whose family has shrunk by over fifty percent in the last thirty years and with Dad's stories ringing in my ears, it was up to me to do something. The aforementioned events and my meetings with Laurent Lapwing and Nige Nightingale galvanised me into action.

I spoke to best mate Bill who had been super busy with nestling care, he agreed something must be done and would support me. His Blackbird family hadn't been affected so much but they were still declining and he wanted to help.

Word got around about the unfeathered plans. Most bird families were suffering from the local habitat changing and all were keen to listen to what I had to say. All bird families in the area attended. There were my close friends Bill, Darcy, Rosie and Jennifer as well as other songbirds: Will Whitethroat, Lottie Linnet and Sal Skylark to name but a few. All were on board as what was going to happen would affect all of them. Then there were unthreatening larger birds like Winnie Woodpigeon and Cerys Collared Dove.

Finally there were the unsavoury, less friendly types like Sam Sparrowhawk, Ron Red-Kite and Meg Magpie amongst others. Normally these chaps would be enemies, or at best, birds we would avoid, however this was a situation that would affect them so they were keen to be involved and swore not to disrupt the meeting for the greater good. If there were less smaller songbirds to hunt and steal eggs from then they would suffer as well. A truce was called, with Rafe Raven providing security in case Sam Sparrowhawk had a rush of bloodlust.

We talked long into the night and discussed what we might be able to do. We did have the benefit of knowing what unfeathereds were saying which was really useful, but we had no way of communicating with them, which wasn't so useful. The spirit was positive and the cooperation was genuine but nobody had a good idea.

Then, just as everyone was growing tired and new ideas were scarce, Darcy Dunnock had a massive brainwave. If anyone was going to come up with a good idea it was her.

"Tom!" she cried loudly, causing some of the more tired birds to rouse from the near-slumber. *"I think I've got it!"*

"You've got what?" I replied.

"An idea!"

"OK, go for it!" Darcy looked excited. *"Listen, I've seen how the unfeathered react when you are singing from the top of the old oak. They love it. I've seen the same reaction when Bill sings in his flutey tone and also with Jennifer when she explodes into song."*

And dare I say it, when I sing I do get admiring glances from the unfeathered. Even though I've been accused of stopping too soon by some birds. Darcy looked accusingly at me as I had questioned her about this before.

She was growing in confidence and it was infectious. *"They really enjoy our singing, every songbird here tonight has a voice and something to offer, or not offer as I will propose. We can use this to our advantage."*

I was listening intently. *"Darcy, you are right, my voice is amazing,"* I said, trying to inject some dry humour to put everyone at ease.

"What are you thinking of doing?"

Darcy replied, ignoring my attempted joke. *"Let's all stop!"*

"Stop what?" said Bill, Rosie, Jennifer and myself together.

"Singing!" said Darcy, as if it was obvious, which I suppose it was.

All the other birds were looking around at each other not knowing what to think or say. Singing is natural for most birds and takes up a lot of time, especially in spring.

Darcy pushed her point. **"Elmdown Wood, the surrounding fields and hedgerows, Little Snuffling. All without our singing, they'd hate it!"**

To be fair we wouldn't like it either, but it was something that might just make them realise they were affecting our futures.

"We can call it 'Tom Song-Thrush and The Big Hush', or something like that," Darcy offered. In honour of our leader with his inspirational thinking!

The assembled birds looked around at each other and at their soon-to-be fledged youngsters. It was the future generations that were going to be more affected more.

A mood of general agreement, indicated by nodding heads, was beginning to spread across the clearing where the meeting was held. Even the unsavouries who can't sing could see the sense in this. For them, no hedgerows, no insects, no flies amounted to no succulent songbirds to eat or to steal eggs from to eat. In an unprecedented show of solidarity, for the task in hand they promised that during the **"Tom Thrush and the big Hush"** period they would concentrate on eating only invertebrates and small mammals.

I was thinking that this could be a brilliant idea. Springtime was over so we didn't need our songs and singing to attract a mate and we had raised our first family. Normally, us songbirds like to continue singing during our second and sometimes third broods but these were special times. We all agreed to keep quiet and, if necessary, forego any further breeding until we had made a difference.

Many songbirds like to sing for other reasons but for as long as it takes they would to stop.

"Darcy, you legend!" I said. *"This is a brilliant idea but I think it will take time for the unfeathered to understand what they're missing. We need tenacity and discipline over a sustained period to pull this off. And a bit of luck! Who's in?"*

There was much positive chirruping and twittering before slowly, one by one, each and every bird raised one wing in agreement.

But I had seen a problem. *"We have one potential issue, chaps. How do we tell the unfeathered that we're silent as a protest due to Farmer Cobbled's and the council's plans to further destroy our habitat? I mean it will be obvious we've stopped singing because they won't hear us anymore but how do we let them know why?"*

Rosie chirped in. *"As you say Tom, it may take a while, but sooner or later they will realise something has happened to cause us to stop singing. They will start asking why. We need to look for an opportunity to communicate with them."*

Jennifer said, *"I agree this is an excellent idea but it won't be easy. As everyone knows I am normally a shy bird but from now on I will make sure I'm seen more often and just sit in silence instead of singing loudly from an unseen place like I normally do."*

I agreed with this idea wholeheartedly and suggested, **"Jennifer is right. From now on we stop singing but we find prominent and conspicuous perches where the unfeathered can see us. We want them to see us not singing, and start questioning why we are quiet."** I then returned to my earlier concern. **"But they still don't know why we've stopped singing?"**

This was a slight flaw in the plan, although Darcy remained positive. **"We have to figure that out as we go"** she said with a flamboyant wing gesture that indicated the will was there and the way will be found.

She finished with. **"As Rosie said, we need to look for an opportunity."**

Bill chirped in, **"Darcy is right, something will turn up, and there's no time like the present, lets get on with it!"**

Rosie and Jennifer nodded in agreement with me not far behind.

"OK!" I said, mustering as much authority as I could and with Dad's tales of thrush decline and my personal promise to his memory swirling around in my head, I announced:

"LET'S DO THIS! LET'S STOP SINGING FROM NOW!"

All the assembled birds left Elmdown Wood in eerie silence.

Chapter thirteen
Late June -
Silence in full view

Making sure we were seen to be silent was important. I decided that for maximum impact I should continue to sit in my normal place at the top of the old oak overlooking Little Snuffling.

So the following morning off I went, I spruced up my feathers and positioned myself at the top of the tree, raised my head, puffed out my chest as if to launch into song and then did nothing.

I started to look around and had noticed a couple of unfeathered had come outside into their garden. They were sitting and drinking their first cup of tea of the day.

I just sat there, head aloft, in silence, hoping they might look at me. I did this for about thirty minutes, which is a long time for me to be silent. I don't think it registered with the unfeathered but I was playing the long game. I surmised that if I did this day after day the unfeathered would become aware that something was missing, especially as they could see me sitting proudly on high.

All the other birds were doing the same, returning to their favourite song perches and surveying the local scene. If they thought it wasn't a prominent enough position they would find another, more conspicuous one.

So all popular places around Little Snuffling had birds sitting in prominent places keeping quiet. Spud Sparrow and Stan Starling who lived in the village and were normally incessant chirpers and tweeters sat with their families in full view and kept silent.

Every footpath, bridleway and track around Elmdown Wood had a conspicuous songster or two in place. Our discipline was great. Morning, noon and night we took up our noticeable places and said nothing. Elmdown Wood, Little Snuffling and the surrounding fields and hedgerows were eerily silent but we didn't care, something bigger was going on and the future of bird life in this area was at stake.

It was a spectacular and unified effort by all songbirds and the camaraderie was strong, the unsavouries were true to their word, all eggs remained in nests and nestlings were fledging successfully in bigger numbers. Even Sam Sparrowhawk resisted snatching unassuming birds out of the air by eating lots of insects and other invertebrates. He even tried seeds and leaves! Sam wished more than anyone for this to be over.

Chapter fourteen
early July -
The Unfeathered take notice

It came to my attention after about a week that a few birds were losing faith in the idea as the unfeathered didn't appear to notice our lack of singing, so I decided to call another meeting to address this situation. This turned out to be a timely meeting as Rosie had just returned from a worm foray in Arthur Ablebody's allotment with some encouraging news.

She had heard Arthur, Mrs Rightly-Soe and Peter Bumbleton, a keen allotmenteer and lifelong Little Snufflinger, chatting about us birds.

"What's happened to all the birdsong?" offered Arthur.

"I was going to ask you!" replied Mrs Rightly-Soe.

Arthur carried on: *"It's incredibly quiet and very strange. I seem to be seeing more birds than normal but they're all as silent as church mice."*

"Maybe we've upset them?" suggested Peter Bumbleton. *"Although how, I have no idea."*

Rosie was listening intently in full view and in silence. Arthur saw her and looked straight at her. *"Little friend, why have you stopped singing?"* He said. *"You are still helping yourself to worms but you're very quiet."*

Rosie, wishing she could speak Humanish, looked back at Arthur for a little bit longer than she normally would and then flew off. She was keen to get back to Elmdown Wood and tell the others what she had heard. Her story seemed to lift the mood and the supposed doubters agreed this was a step in the right direction.

Then I had an idea. It was now apparent that the unfeathered were beginning to notice our silence, so I thought it would be a good idea to do some more focussed and furtive listening to unfeathered conversations.

I asked Darcy to spend some time in the hedges around Farmer Cobbled's house to see if she could hear anything interesting about his plans or whether he had noticed our bird silence.

Jennifer was assigned a similar task outside the village hall by the bus stop where the unfeathered congregate a couple of times a day. This might be a good place to listen to general chit-chat.

Bill spoke to Buffy and told her I needed him. Could he spend a little time helping out? She understood, and said yes. So I assigned Bill a spot called the Four Trees where two bridleways and footpaths cross. Bill was there like a shot, glad to be helping out by putting his nuclear-powered ears to good use.

I placed myself in a tree overlooking the centre of Little Snuffling. Babblybrook, the village stream, ran through here and unfeathered would often sit and chat on the wooden benches next to the water.

Other songbirds were told to pay careful attention to any unfeathered talk as they walked through Elmdown Wood or were seen on the bridleways, footpaths and tracks in the area.

It had been over a week now and I was starting to have big withdrawal symptoms from singing. As you know I love it and also enjoy others listening, so this was hard. I tried singing into a hole in the ground to dampen the sound, but …

A – I looked ridiculous.

B – My voice was deafening in such a confined space.

So I stopped doing that and steeled myself for more silence.

Jennifer was also struggling. For one who is used to exploding into song at any given moment, silence was not normal.

It was slightly different for Darcy, normally she stops short of a full song anyway so this was like a respite for her. But deep down, as a songbird, it was still tough.

Then for the first time one day I noticed a couple of unfeathered looking at me and pointing. I was stony silent, of course. I strained my ears. *"Why is that Song-Thrush not singing any more?"* said one. Her name was Jane Jaunter. *"He's got such a great voice and I am starting to miss it!"*

Her friend agreed. *"Do you know what, it's really strange. I was out walking recently and saw lots of lovely birds, more than normal I would hasten to add. I walked all around the local fields and over towards the White Clumpys, but every bird was silent. Even the annoying rooks and crows. It's weird."*

"I agree," Jane Jaunter replied, *"it's not normal, I don't like it. I'm also seeing more birds than normal but none of them are singing. If only we could ask them what the problem was."*

This got me going. As I couldn't speak Humanish, the best I could do was make a physical gesture. So I started hopping up and down, which must have looked a bit strange as I'm normally a singer not a dancer. I was thinking the same thing as Jane Jaunter. If only we could tell you why we weren't singing!

The two unfeathered were now totally confused. Their reliable Song-Thrush songster was now trying to dance?

Chapter fifteen

mid July -
Ollie and Lucy
miss the singing

A couple more days of the birdsong silence had elapsed when more good news filtered in. Darcy had been down at Farmer Cobbled's farmhouse for a few days, and had positioned herself opposite the front door in a privet hedge. Normally quite shy, today she was brazenly on top of the hedge staring at the front door. There were lots of comings and goings, particularly from Farmer Cobbled's two young kids. Their names were Ollie, who was ten years old, and his sister Lucy, who was eight years old, and they talked constantly. Nothing much of interest was said on the first day apart from **"that hedge sparrow doesn't normally sit there"** from Ollie.

Darcy doesn't like being called a hedge sparrow but was determined to stay in position and say nothing. On day two she positioned herself nearer the garden where Ollie and Lucy played a lot. Sure enough they appeared, with their mum this time, and saw her immediately. Darcy was treated to some crucial information.

Ollie said, **"Mum, I saw that bird outside the front door yesterday and I don't normally, now she is here. And she's very quiet, Dad told me hedge sparrows have a very nice song. Why isn't she singing?"**

Darcy overlooked the "sparrow" reference again for the greater good.

Ollie continued: ***"And my Song-Thrush who sings to me to sleep in the evening and wakes me in the morning isn't singing either and I miss hearing him!"***

Sister Lucy piped up, ***"Yes mum, and why is my pigeon friend also quiet? I miss her coo-cooing as well. When will they start again?"***

Mrs Cobbled was already aware of the bird silence and didn't know how to answer. ***"Well kids, I don't know why they've stopped singing, I'm the same, I miss their songs as much as you do. I wish I knew the answer."***

She was looking at Darcy when saying this. Darcy was tight-beaked in terms of singing but flapped her wings a little and was wishing she could speak Humanish to explain to Mrs Cobbled. She eventually flew back to Elmdown Wood, found me and related the story. It seemed to me that our silence was starting to have an affect on the unfeathered. I passed Darcy's story on to the other birds.

Like me, lots of birds enjoyed singing and were going stir crazy not being able to. Bertie Blackcap who also missed singing tried the **"stick your head in a hole"** trick, others tried singing under the waters of Babblybrook, which is not recommended unless you want to choke or possibly drown. And some tried flying really high so as not to be heard. The trouble with flying high was by the time they arrived at an altitude out of earshot they were too tired to sing.

So Darcy's news was most welcome as it seemed we were having an effect, and what we were doing was not a waste of time. Things were starting to happen in terms of the unfeathered noticing and expressing sadness at our stone cold silence.

The next problem, once we had their attention, was how to tell them why we weren't singing.

Chapter sixteen
late July -
Enter Osbert the Old

Little Tim, Ben, Josie and Jill were almost fledged now. As I watched them taking their first tentative looks over the side of the nest I felt a rush of pride coupled with apprehension.

Soon they would be out in Elmdown Wood and a big, wide world beyond planning their lives … if there was enough food and enough nesting sites still around. This thought frightened me. I must make the unfeathered understand our worries.

Time for another meeting. All birds loyally turned up, each reporting on their week's silences. More stories were filtering through of unfeathered commenting on the lack of birdsong. We needed a plan to take this to the next stage.

The meeting was in full flow with everyone trying to contribute an idea at the same time. It got a bit chaotic and I was struggling to keep order. All of a sudden from a high branch came the deep and authoritative voice of Osbert the Old. Osbert is the Tawny owl who ceremonially joined Bella and me. He had been in Elmdown Wood all his life and was the oldest and wisest bird I know.

He said, *"I have an idea. I've lived in and around Elmdown Wood since I was an owlet and that was a long time ago. No bird know the ways of the unfeathered better than I."*

Osbert the Old was a respected fellow and immediately gained all the birds' attention, apart from Benny Blue-Tit and his prankster mates, who continued hopping about and fighting with each other.

Rafe Raven, our security, was forced to have a word in their ears. They shut up pretty quickly after that.

Osbert continued in a serious tone. *"The village hall in Little Snuffling is the place where the unfeathered have meetings and make decisions about village affairs. There will be a meeting soon between Farmer Cobbled and the local government to decide when and exactly what is going to happen to the surrounding fields and hedgerows."*

Everyone was listening. Osbert went on. *"We need to find out when the meeting is, also the council make their decisions in the village hall as well. They will hold a meeting to discuss when to mow the verges and destroy the flowers and grasses, so we also need to know when that meeting is."*

I was listening intently, then spoke. *"May I ask what for, Osbert?"* I said, trying not to sound negative.

"What for!" Osbert almost shouted. *"What for!"* he said again. *"We need to be there … we ALL need to be there, well at least outside the hall! Us birds need to make our presence felt, the unfeathered don't like our silence and are looking for reasons why we are behaving this way."* He went on, his voice raised. *"We need to be there and make a plan about what to do. It may be we have no idea until the moment comes, so we need to be spontaneous. The important thing is we are there, all the different birds and their families if your nestlings have fledged."*

I was a little shaken by Osbert the Old's loud and passionate response. *"OK, we'll be there,"* I said, looking around at the assembled feathered friends and other birds who, although feathered, could not be described as friends. I was referring to the unsavouries who were united with us in this cause.

I repeated what we were going to try to do. **"If we can find out when the meetings are we will all be there and I will give some thought to an effective plan."**

Osbert spoke again and motivated us with some wise words. **"You are brave and determined little songbirds and are doing an amazing job by staying silent. It's beginning to work, the unfeathered are starting to miss your voices. You need to keep going a little longer, it will give us more of a chance to help ourselves."** He said inspirationally and offered more advice: **"Don't forget, all unfeathered conversations are worth listening to.**

It may not be just outside the village hall where information about meeting dates can be gathered."

He had a point. It could crop up in their conversations anywhere. I talked to my close friends after the meeting. It was decided that Bill, Rosie, Darcy and Jennifer would rotate their time outside the village hall and bus stop. When not at the village hall they would frequent their normal vantage points: Rosie – allotment, Darcy – Farmer Cobbled's farmhouse, and Bill – Four Trees where two bridleways and two footpaths converge. A favourite spot for unfeathered walkers.

Jennifer Wrennifer, when relieved of village hall duties, would skulk around the local hedges and bushes trying to earwig unfeathered conversations.

Other birds continued in their spots. Eve Yellowhammer mid hedgeline between Elmdown Wood and Little Snuffling, Bertie Blackcap at the entrance to Elmdown Wood and Cerys Collared Dove perched on a wire at the end of the main street in the village.

I would continue at the top of old oak and by Babblybrook, spending a few hours at each.

All of us were in full view and tight beaked. The silence was going to become deafening to the unfeathered.

Chapter seventeen
early August - Why no singing anymore?

Ollie and Lucy Cobbled were getting more upset, for the same reasons as before. Lucy, because she couldn't hear Winnie Woodpigeon cooing outside her bedroom window anymore and Ollie because I (Tom Song-Thrush) wasn't waking him up in the morning and singing him to sleep in the evening.

In addition Mrs Rightly-Soe was not enjoying her early morning cup of tea in the garden surrounded by silent birds.

An interesting and encouraging conversation was heard by Bertie Blackcap near Elmdown Wood. Bertie was positioned at the top of an elder tree on the corner of the wood when two unfeathered appeared. They were out walking their dogs, and started chatting to one another.

One man said, **"The countryside is so quiet without the birds singing, it's driving me crazy, this is not right. We need to find out what the problem is and fix it but where do we start!"**

His mate replied, **"Maybe they know something we don't. Maybe the end of the world is coming and they're too sad to sing?"**

The first man looked sideways at him as if he was a bit mad but hoped it wasn't true!

Chapter eighteen
mid August -
Old friends disagree

One morning, I was doing my normal, silent puffing out of my chest in full view of the benches by Babblybrook when I got the information we required.

Farmer Cobbled pulled up in his Land Rover. He had seen Arthur Ablebody walking by and wanted to talk to him. They had been friends for a long time. **"Arthur!"** he called, **"Are you coming to the meeting next week about my plans for the farm?"**

Meeting next week! I heard this and tuned in immediately.

Arthur replied, **"I'll be there, but I have to say I'm not in agreement with all this hedge removal and mowing of field edges."**

This was turning into my kind of conversation.

Arthur continued. **"Hugh, it's not good for the environment and the wildlife. I wouldn't be surprised if it's the reason the birds have stopped singing."**

As he said this he shifted awkwardly from side to side and looked away from Hugh seemingly to avoid eye contact. I think he was aware he sounded ridiculous but had the courage to say what he felt which was extremely helpful to us!

"Oh!" Farmer Cobbled replied. *"I'm sorry you don't agree, but it makes financial sense for me to do this and as for your theory about the birds … well, thats pie in the sky"*

"It's not all about money, Hugh." Arthur replied.

Farmer Cobbled looked a bit deflated and disappointed that his friend didn't agree with him but managed to say. *"See you next Wednesday at seven thirty p.m. then, they are combining my plans with the council meeting about grass verge mowing on the same evening."* He looked at Arthur. *"I suppose you don't agree with that as well?"* He sounded rather sheepish.

Arthur had heard about this and was again not happy. *"I'm afraid I don't, Hugh. What's happening? We're ruining our lovely countryside! You for money, and a few small-minded village folk for their selfish ideals. Sterile, short and frankly soulless grass verges are not a good thing."* Flustered, Arthur finished with, *"I will object!"*

Farmer Cobbled was taken aback. His mate for many years had fallen out with him. He started to think.

I was excited. I had all the information about the meetings and was really pleased to hear that they were both on the same evening. This was our opportunity! When I thought about this more I wasn't entirely sure what the **"opportunity"** was but I did know that we should be there in case one should arise.

I called a meeting and explained what I had heard said between Hugh Cobbled and Arthur Ablebody.

Bill had also heard a conversation whilst at the Four Trees amongst a group of three unfeathered dog walkers. They were also talking about the village meeting and how they didn't agree with the council's and Farmer Cobbled's proposals. One of them actually looked at Bill and spoke: *"See that Blackbird up there? He and all other birds haven't sung for weeks, I wouldn't be surprised if they somehow know what is going on and this is a protest!"*

Bill was flabbergasted when he heard this and started jumping up and down in the same way I did on hearing a similar conversation.

The second unfeathered commented, *"I think you've hit a sore point, look at him jumping up and down, that's not normal Blackbird behaviour! Maybe they can understand what we are saying?"*

Bill was going apoplectic now which was amusing the unfeathered.

"You're bonkers!" said the third unfeathered to his two friends, *"Birds understanding what we say! You're both off your rockers!"*

On hearing these two encouraging pieces of news, our fellow birds' spirits were lifted.

Chapter nineteen
Late August – Fun at the farm

The next day I flew over to Farmer Cobbled as I was curious how Ollie and Lucy were getting on without their favourite birds serenading them. I perched in my normal position in the old oak and waited for some activity. I was soon blessed with some enlightening information. Farmer and Mrs Cobbled appeared out of the front door of the house closely followed by Ollie and Lucy who were clearly not in a good mood.

Mrs Cobbled was trying to explain to her husband why this was. *"Hugh"* She said. *"This is not a happy house at the moment. Ollie and Lucy are really upset that the birds have stopped singing. Ollie misses that lovely Song-Thrush singing to him both morning and evening and Lucy is the same with the cooing Woodpigeon who sits outside her window. They are still around and the kids see them but they are silent. It's weird."*

I wasn't aware that Ollie used to watch and listen to me singing in the old oak, but was happy he liked the show and it gave me a little more confidence in our plan.

Farmer Cobbled looked bemused and a little confused. *"I know Libby,"* he said. *"I've lived here all my life and never have the birds behaved like this before. It's nearly two months now and I'm starting to think I have something to do with it."*

"What do you mean?" replied Libby.

"Well you know .... My plans to make the fields bigger by removing all the hedgerows and mowing the field edges.... It seemed like a good idea at the time as we can make more money but I know as a country person that it will affect the birds nesting and feeding habitat.. And I accidentally killed a Song-Thrush with my car the other day. I'm starting to get paranoid about this situation!"

Libby replied. *"You are, aren't you, I dont think it's your fault but then I have no idea what's causing this?"*

Hearing Farmer Cobbled mention Dad was painful but I was heartened by the fact that he was thinking about our plight.

Little Snuffling's tidy brigade, Johnnie Cotton and Dick Moss, couldn't wait for the meeting so they could find out exactly when the overgrown verges and messy wildflowers were going to be sorted out.

"It won't be long," said Johnnie, *"and all will be ship-shape in the village and my very expensive house will look less like it's surrounded by poverty and people who don't care about tidiness."*

"I miss the birds singing," said Dick innocently.

"You miss the birds singing?" said Johnnie. *"Having a break from that constant 'cawing' from the rooks is welcome in my book."*

"Good point," said Dick, *"they are noisy."*

Chapter twenty

early September - Councillor Dunwell

Now is a good time to introduce Councillor Dunwell. He has an important part in our story.

I wasn't aware of the following at the time but my super inquisitive friend Rosie Robin had been spending lots of time at the allotment listening to Arthur Ablebody, Mrs Rightly-Soe and Peter Bumbleton talking. Recently they had been discussing in some detail the new councillor at Little Snuffling and some of the problems he had been facing.

She had gleaned lots of information during her extended worm forays that she was eager to share with me. According to what she had heard it seemed Councillor Dunwell was a good man and time-honoured servant of district civic and community matters. He had worked his way up from office junior thirty-five years ago to become head of Roads and Environment at Bulgingham Council. Bulgingham is the biggest town in the area and somewhere we Elmdown Wood birds don't visit often.

A year ago Councillor Dunwell decided to leave the bigger council and move to Little Snuffling. He was approaching fifty-five years old and thought it was time to take a less stressful job in a smaller local council. The Little Snuffling job came along and looked ideal. The salary wasn't as good but that wasn't important.

The new role included responsibility for grass and wildflower verge maintenance, local planning permissions, street maintenance and all environmental issues. As you can see, he was someone who could help us!

Councillor Dunwell was looking forward to his new role and was settling in nicely. The first six months were very refreshing and at a much slower pace than at the town council. He was getting to know the village and its people and thoroughly enjoyed accepting all the invitations to local charity events and committee meetings where he was treated like royalty.

It was just as he was relaxing into this different pace of life and starting to understand village life that the complaints started to come in. He wasn't prepared for what was to happen next. In fact, I don't think any of us were! He started getting many phone calls and letters from disgruntled unfeathered villagers. The complaints were all saying the same thing (apart from Ms Dingled who had written to say she had a problem with a bee in her bonnet).

... and that was ...

"We don't think Farmer Cobbled should destroy all these hedgerows and mow all the field edges."

Or ...

"We definitely don't think you (the council) should be mowing the grass and wildflower verges. It's bad for the environment. Return the new mowers immediately!"

Or ...

"The birds have been silent for two months now, we think they are protesting about the plans for the fields and hedgerows and verges."

This complaint got Councillor Dunwell thinking and he decided to investigate further.

There were a few unfeathered who were in favour of Farmer Cobbled and the "tidying" plans. They emailed Councillor Dunwell saying things like *"It's about time we sorted out our untidy verges, to help curb the swarms of insects flying about the place"*, and *"Good on Farmer Cobbled, trying to do his bit to feed our country by making his fields bigger!"*

But generally there was a groundswell of feeling amongst the unfeathered that tampering with the environment and local habitat was wrong and that somehow it might be linked to the silence of us birds. The unfeathered were starting to understand the bigger picture, I was cock-a-hoop! We were slowly moving in the right direction. This **"Tom Song-Thrush and the Big Hush"** campaign of ours was becoming a hot topic of conversation.

Chapter twenty one
early September -
The silence goes viral

Recently, Rosie was flying back to Elmdown Wood from Little Snuffling. She had been on one of her favourite perches, the old metal road sign in the middle of the village. She would sit there and try to attract the attention of the unfeathered walking by. She was dying to serenade the village with her melodious, flutey voice but knew she couldn't.

On her way home to Elmdown Wood she spotted two unfeathered chatting not far from Charles Chaffinch's vineyard. Being the inquisitive bird she is, Rosie decided this was another opportunity to stop and display herself in silence. She found a nice branch in an Elder tree just in front of the two unfeathered. They were waving their arms about and laughing. Rosie listened.

"It's unbelievable", said the first one. *"Little Snuffling is becoming quite famous!"*

"I know!" The second replied. *"And all because the birds have decided to stop singing. I have to say it is pretty strange but didn't think it would become of national interest!"*

Rosie was intrigued now and her heightened powers of inquisitiveness were being put to good use.

"That's social media for you," said the first one. *"A tweet here and a tweet there and before you know it, it's gone viral."*

Rosie was a bit lost now. Who's tweeting? She thought. Birds tweet, not the unfeathered?

She thought some more. Who's famous?

She listened again.

The second unfeathered replied, *"Yes, it's all over Instagram and Facebook as well."*

"Instagram", **"Facebook"**, this was a part of the Humanish language Rosie didn't understand.

The second one continued, *"I met a chap in the shop the other day. He was visiting the area just to see lots of birds sitting in silence. The birds are becoming celebrities and they don't even know it."*

We know it now! Thought a wide-eyed Rosie.

The first one spoke and pointed at Rosie. *"Look at that robin up there. Quiet as a quiet thing in quiet town. I used to like listening to her or one of her buddies singing."*

Rosie was happy the unfeathered liked her singing and was indeed quiet. But her mind was racing ...

... **Is this a good thing?**

... **Is this a bad thing?**

... **Will there be too many unfeathered coming to see us?**

... **Will this help us with our plan?**

Rosie had lots of thoughts but no conclusions. She thought she better get back and tell me as soon as possible.

The first unfeathered gestured to Rosie. ***"Hey, Mrs silent robin sitting in the tree, what's your game? The villagers, me included, don't understand your unusual behaviour and now, nor does the whole country!"***

Rosie looked at them, flapped a bit, puffed out her red breast and then flew back to the other side of Elmdown Wood to tell me what she'd heard. She found me with little Tim, Ben, Josie and Jill. I was trying to explain to them what happens in springtime.

Little Tim and Ben were getting excited about being able to sing like their dad and using this talent to meet a lady Song-Thrush.

I was thinking, Boys, there's no hurry! But I didn't say it. This was the advice I gave myself at little Tim and Ben's age, but then Bella came along and blew my socks off – if birds wore socks – and look where we are now.

On the flip side, the girls quite liked the idea of being serenaded by a handsome male Song-Thrush. For a moment, I hoped they would meet a nice one like me.

Rosie flew down next to us. ***"Excuse me kids, can I have a word with your dad?"***

"Sure thing," they chorused, mildly relieved to move away from Dad's life lesson.

"Tom," Rosie said, *"I've just overheard a couple of unfeathered chatting and guess what?"* She had a slightly bewildered look on her face, which meant to me that Rosie's inquisitive nature had uncovered something noteworthy.

I could've tried guessing but wasn't really in the mood as it sounded like she had some important information which I needed to hear forthwith. *"Just tell me please?"*

Rosie looked mildly offended but quickly realised that now wasn't the time to be sensitive. *"It seems our* **"Tom Song-Thrush and the Big Hush"** *campaign is becoming known throughout the country,"* she said.

"What?"

"I know!" said Rosie, *"Can you believe it, unfeathered are travelling here from far and wide to see us birds not singing. I got a bit lost then, but apparently the unfeathered can tweet, they do it a lot and they use it as a way of communicating."*

I was getting a bit lost now as well. *"I've never heard them tweet!"* I said.

"Me neither!" said Rosie, *"But that's not important. What is important is that lots and lots of the unfeathered are aware of our unusual behaviour."* She went on to talk about **"Instagram"** and **"Facebook"** and again I had no idea what she was talking about. From what she said, it seems these are Humanish words for things they use to communicate with. I thought, maybe we could have **"Facebeak"**?

I'll hold that thought for later.

"Tom," said Rosie, **"this is a good thing, right?"**

I thought for a moment and then said **"Yes, I think it is,"** with an enthusiasm that grew in me as I spoke. It had been two months now without any singing birds and the countryside felt strange. The majority of the unfeathered didn't like the silence. It was a popular topic of conversation and various theories were being discussed.

One theory was that there had been an increase in radioactivity from the sun which was making the birds ill, another was that Babblybrook was contaminated and the birds were being poisoned. Both of these seemed a bit far-fetched.

Then there was a third theory, which was equally far-fetched! A couple of astute unfeathered had realised the birdsong silence coincided with the announcement of the aforementioned plans. **"We think they can understand what we say and don't like Farmer Cobbled's and the council's plans,"** they suggested. **"The birds are protesting."**

The unfeathered in question had no idea how hard they had hit the nail on the head! In truth, nobody knew the cause but the unfeathered agreed they needed to find a solution as country life would be unbearable without the birds singing.

The village hall meeting was approaching and we needed to have a plan. With Rosie's recent news it seems that some unfeathered were linking the planned environmental changes to our stone cold silence. Which was only good news for us.

Chapter twenty two

mid September - Waiting for an opportunity

I called another bird meeting.

It was tricky to think of an effective strategy. We could just turn up and all sing loudly together outside the hall as the meeting was about to get started. That would cause a stir and quite possibly the unfeathered could deduce from the cacophony that we were there to complain. But I thought we needed to be a bit more subtle and the timing needed to be more appropriate. I discussed this with Osbert the Old and he agreed we should bide our time.

"Have faith," he said. *"An opportunity will present itself."*

I made a point of speaking to all the birds individually, from Ken Corn-Bunting to Bram Buzzard and Rob Rook to Philomena Pheasant, I told them what we were to do. I explained that Osbert the Old believed there would come a time during the proceedings when we could make our point.

As I knew from my previous earwigging of conversations at the farm, Farmer Cobbled was having a hard time and was thinking deeply about his situation. He loved his family and didn't like to see them upset. Ollie and Lucy were still unhappy about Winnie and me not singing for them. He started asking himself, was he doing the right thing by changing the environment on the farm? Life was pretty good and he hoped one day Ollie would take over. However the way things were at the moment, Ollie would be on the first train out of here, if he was old enough and there was a station at Little Snuffling. Hugh Cobbled was a nice person and no fool. He knew from what he had read that increased field sizes, more use of fertiliser, fewer hedgerows and an absence of untidy field edges did have an effect on us birds.

As I said earlier, Councillor Dunwell, who was surprised at the level of objections to the plans, was intrigued by the theory that we understood them. He started to consider the bird silence more seriously.

Chapter twenty three

Late September - Judgement Day!

Village hall meeting day was upon us.

All the birds were excited and apprehensive and no one more so than me! I started this whole thing to help my fellow Song-Thrushes and other birds, and felt very responsible. Darcy's idea was a game changer but my concerns, passion and drive got the ball rolling and I had assumed leadership. As the leader and figurehead I was the one in charge of decision-making on the day.

No pressure then. I had no idea what I was going to do but just trusted Osbert the Old when he said *"an opportunity will present itself"*.

We all gathered in Elmdown Wood, over forty species of feathered friends and previous foes, all with one purpose. Today was the culmination of over two months of hard work and sacrifice, and I was a bit emotional.

I made a short speech.

"Today is hugely important if we birds are to continue living happily in these woods, hedgerows and fields. If the unfeathered farmer and council do what they are planning we will have to move away, or worse, we will perish. If they get away with it this time, they will do so again and again, and our families will all die."

When we all rose up into the air as one and started the short flight across two of Farmer Cobbled's fields to Little Snuffling and the village hall, I had a tear in my eye.

Big birds next to small birds, victims and predators side by side in a show of solidarity. Unsavouries like Kevin Kestrel, Sam Sparrowhawk and Ronnie Red-Kite were flying side by side with Sal Skylark and Pascale Partridge who would normally be their lunch! It was an impressive sight with Osbert the Old flying alongside me at the head of the aerial cavalcade. I thought about how Dad would have felt.

The meeting was due to start at seven thirty p.m. and we arrived at seven. A few early unfeathered were filtering into the village hall when we swooped gently down and alighted on the old red brick wall opposite. Babblybrook was very narrow there and ran in front of the wall. The village hall was an old, thatched building that had been there hundreds of years and had held countless numbers of civic meetings but none quite like this!

The mass flight was all done in complete silence apart from the fluttering, flapping and whirring of many wings. The effect on the unfeathered witnesses was profound. They stared open-mouthed as we appeared, then quickly shuffled into the hall chattering nervously to each other with worried looks. I wish I could tell them they had nothing to fear as we weren't going to cause any trouble. This was a peaceful but necessary protest.

Farmer Cobbled and his family arrived. I thought having Ollie and Lucy there was a good thing.

Then Councillor Dunwell and his wife, who was an ardent dog-walker and pro-birdsong.

Others supporting us – Arthur Ablebody, Mrs-Rightly-Soe and Peter Bumbleton – arrived next. Arthur looked across at us and smiled. He caught Rosie's eye, **"There's my little robin friend over there!"** he exclaimed, smiling and waving at Rosie. Rosie didn't wave back but she looked directly at Arthur and flapped her wings. **"This is unprecedented,"** said Arthur. **"These birds know what the meeting is about!"**

Mrs Rightly-Soe agreed. **"Yes Arthur, I think you're right. How exciting! Seeing them all on the wall outside was surreal, I wonder what they are going to do?"**

Also arriving were the whole of Little Snuffling's dog-walking brigade, whose conversations were overheard by us birds earlier. They missed our singing as well.

Then there was Johnnie Cotton and Dick Moss, who couldn't wait to cut all the grass everywhere. They glanced briefly across at the birds but failed to appreciate that this was unusual. Birds and birdsong did not register on Johnnie's radar although Dick did mention that he was missing it. Johnnie's life, especially, revolved around working and competing with his unfeathered neighbours in a never ending competition of who has the biggest and best one. Johnnie and Dick had also brought with them friends who normally would be their competitors. They had convinced them that mown grass verges and an absence of wildflowers were the way forward.

Other villagers turned up in their droves, and the village hall was full. As I mentioned earlier, Little Snuffling had become a talking point throughout the country due to extensive tweeting and twittering, by the unfeathered, not us! Newspaper headlines read **"The village where birds don't sing"**, or **"Silent Birdville"**, or **"The silence of the birds"**. There was even a TV crew from a national station. They couldn't believe their luck when they saw us all assembled together. **What great TV!**

Everyone was talking about us birds on the wall opposite. Later arrivals had not been treated to the spectacle of us initially flying in but we were still a formidable sight. To the unfeathered it seemed supernatural.

As I was the leader, I thought it appropriate to watch proceedings first hand (or should I say, first claw?), after all I would understand everything that was going to be said. With this in mind, during the previous week I had scoped out the hall for a place to observe. I found an open skylight in the apex of the roof. Just inside was a metal bar that was the perfect perch for reconnaissance. I just hoped the skylight wasn't closed on the day!

Luckily it was still open and I positioned myself just inside on the window latch and waited for the meeting to start. I wasn't sure what I was going to do, but I thought it important to be there and listen to what was being said so I could react if I needed to. The other birds were briefed to sit tight and keep quiet. If something important happened I could be back with them in seconds, to instruct as necessary.

In terms of officialdom, in the hall was Councillor Dunwell, who was Chairman, and basically the person who made the decisions. Assisting him was Mr Prods, a young unfeathered who took notes and organised Councillor Dunwell's business life. Alongside them on one side was Trevor Dredging, the environment manager who looked at any impact the aforementioned plans would have, and also a lady called Flossie Beetledown who was responsible for the local roads and for the wildflower and grass verges.

Just as the meeting was about to start, Little Ollie Cobbled shouted out: ***"There's my Song-Thrush!"*** and pointed at me in the skylight and waved. ***"Hi Tom!"***

How did he know my name, I thought? I sat still and didn't move, this wasn't going to faze me.

Ollie said sadly, *"Dad this is my favourite bird, he used to wake me in the mornings and sing me to sleep in the evenings but he doesn't any more. Dad, why?"*

Everyone in the village hall was waiting and listening intently for the meeting to start. Councillor Dunwell stepped in. *"Quiet please, everybody."*

Farmer Cobbled shushed Ollie and promised to sort out the lack of singing.

Councillor Dunwell then spoke. *"Thank you all for coming and aren't there a lot of you! We are going to have a busy evening and something tells me tonight is going to be memorable."* He didn't elaborate on why, but I think most unfeathered agreed. Little Snuffling had a unique problem that tonight it may or may not fix.

"I am going to address the elephant in the room immediately," Councillor Dunwell said. *"Does anyone have any idea why all the birds in the neighbourhood are sitting on the wall outside and more importantly why have they not sung for many weeks?"*

This was a rhetorical question and probably should have stayed in his head, as people started to raise their hands and shout things out. *"Stop!"* said Councillor Dunwell. *"We will get back to that later. We have two items on the agenda this evening. Both of them proved to be highly emotive."*

In the meantime I had regained my composure after being put in the spotlight by Ollie Cobbled.

"Item one:" said the councillor. ***"Increasing field sizes and mowing field edges on Farmer Cobbled's farm. Reasons for: higher yield, more economical harvesting. Impact on countryside: loss of hedgerows, loss of wild and natural field edges, higher use of damaging fertilizer, bird habitat destruction, invertebrate habitat destruction, encouraging a sterile environment**.*

Item two: Mowing the grass and wildflower verges in and around Little Snuffling. Reasons for: tidiness, and safety, as natural growing verges could obscure unfeathered drivers' view."

I couldn't argue with this but mowing need only be a short distance each side of a junction.

The councillor continued. ***"Impact on countryside: loss of pretty and colourful wildflowers, and in turn loss of pollinators such as bees and wasps, the loss of invertebrate habitat, reduction in food for birds and wildlife. Let us start with item one. Mr Cobbled, can you tell us more about your plans please?"***

"Thank you, Councillor," said Farmer Cobbled, Ollie's plea to help bring my singing back still ringing in his ears. ***"If I'm honest, I would simply like to make my field bigger so I can feed more people and make more money. The government will help me do this so it's really a no-brainer."***

His friend Arthur Ablebody raised his hand. Hugh Cobbled expected this.

"Go ahead," said Councillor Dunwell.

"Hugh, you know the damage this will do to the wildlife environment, especially the birds. We are in unprecedented

times and all the signs are that something like this should not be done." He went on, *"I'm sure you can still have a good life with the things the way they are."* Arthur was getting going now. *"Call me a crazy old man but I am convinced the birds know what is happening here and that is why they are all outside on the wall. They are protesting!"*

Ollie called out again: *"Mr Ablebody, can you make them sing again please!"*

Farmer Cobbled shushed Ollie again and mumbled. *"I've said my piece."*

"That is an interesting idea about the birds," said Councillor Dunwell in a genuinely interested way. *"Would anyone else like to say anything?"*

An unfeathered man stood up. *"I was up at Elmdown Wood the other day talking to my friend about the birds and I agreed with the previous gentleman who spoke. It looks like they are protesting about something. I remember pointing at a blackbird in a tree. He looked very agitated and was jumping up and down in silence, I am sure he knew what I was saying."*

"It's funny you should say that,"' said another unfeathered: it was Jane Jaunter. *"I was by Babblybrook in the centre of the village the other day talking to a friend. We were discussing the silence of the birds and how it didn't feel right. As we were doing so I pointed at a handsome Song-Thrush in a bush next to the water."*

She remembered me! Handsome? I did blush a little at this point. Jane had obviously missed Ollie's earlier outburst and didn't realise I was there, as she continued: *"I looked at him and said to my* friend, wouldn't it be great if they could just tell us what the problem is? It was then the Song-Thrush started jumping up and down frantically, like he was trying to dance.* Then she said to the other unfeathered man, *"Similar behaviour to your blackbird by the sound of things."*

  "This is all very interesting," said Councillor Dunwell. *"Some of it sounds a little far-fetched, I have to say, but I'm not dismissing your testimonies."*

At this point there was noone in the hall who was vocally pro Farmer Cobbled's plans, which was excellent. From my vantage point I was quietly hopeful things were going in the right direction.

Councillor Dunwell then asked Trevor Dredging, the environmental manager, if he wanted to say anything. What happened next was a bit of a hammer blow. Mr Dredging stood up and without hesitation said, *"I fully support Farmer Cobbled's plans. I have looked into the environmental impact and it will be minimal."* He had clearly made his mind up before he had entered the building and wasn't interested in saying too much. Then in an attempt to ingratiate himself with higher powers he went on: *"What's more the government thinks it's a good idea so it should go ahead."*

I was instantly deflated. Did this mean it was all over and we were going to get further habitat destroyed? I was shocked. Arthur Ablebody, Mrs Rightly-Soe, Peter Bumbletone and all the dog walkers were also looking shocked and sad.

  "This is pants!" shouted Arthur.

  "Absolute tosh!" shouted the mild-mannered Peter.

"Does this mean the birds will never sing again?" screamed young Ollie.

The dog walkers were about to walk out in disgust but decided to stay until the end to see what the birds would do.

"Quiet please," said Councillor Dunwell, *"and thank you for your input, Mr Dredging."* He did not sound convinced.

I was ready to fly back to the others and tell them it was all over but something told me to stay on. Osbert the Old's words, *"an opportunity will present itself"* were still in my head so I stayed put. It's not over until the fat bird sings as they say and she hadn't sung yet.

Councillor Dunwell continued. *"Item number two is about mowing the many grass verges in and around Little Snuffling."* He looked at the roads and verges maintenance manager. *"Miss Beetledown, please tell us of your plans."*

Miss Beetledown didn't look very old and certainly didn't look particularly confident as she stood up to answer Councillor Dunwell. *"Well,"* she said nervously, *"it is my job to ensure that the grass verges on the roadside are kept tidy. I have driven around the area and sure enough there are some places where the grass is long."* Then in an open admission of what she was feeling, *"But the flowers are very pretty."*

This was good, I thought. It seems as if she may be on our side. My mood lifted a little.

"However," she continued, starting to gain confidence now she had spoken a little, *"there is the safety consideration. Car drivers' views are obscured at junctions by unmowed verges and we cannot have that."*

I was listening and had to concede this point but hoped she wouldn't finish there. But bingo! Miss Beetledown wasn't finished.

"On the other side of the coin," she said, *"I believe it is important to keep the verges natural and allow the wildflowers and grasses to grow to help sustain the local wildlife. I have a degree in conservation and ecology and understand how the loss of these habitats impacts the food chain from bees to buzzards and everything in between."*

This was music to my ears! I flapped a bit in appreciation and several unfeathered looked up at me.

Miss Beetledown was now getting into the speech, as her plan was taking shape. *"I propose we can keep the roads safe and preserve the natural habitat with some careful thought. We will survey the dangerous junctions and mow both sides to achieve line of sight for drivers. We will then stop and leave the rest of the verges untouched. I predict only about ten percent of local verges would have to be mown."*

I could hardly contain myself and started jumping up and down again, my way of trying to communicate with the unfeathered. As this was happening Jane Jaunter, who'd seen my previous jumping-up-and-down fit, stood up and pointed at me. *"Look at that lovely Song-Thrush up there, he's jumping up and down again!"* She was getting quite excited. *"He knows what we're saying, what a clever thrush!"*

Miss Beetledown and Councillor Dunwell also looked up at me "dancing". I decided after hearing Jane Jaunter's outburst that jumping up and down was helping, so I really went for it and nearly fell off my perch.

"Objection!" a voice shouted out. It was Johnnie Cotton, losing his temper. *"This is a load of old nonsense! Cut the verges and tidy this village up! I spent an awful lot of money on my really big and impressive house. When I drive out of the gates, what do I see? Long grass and higgledy-piggledy flowers all over the place."*

Johnnie's mate Dick Moss also stood up. *"Yes, long flowers and higgledy-piggledy grass,"* he said, getting it wrong.

Johnnie glared at Dick. *"It's dangerous as well!"* he said. *"Get it all mown now!"*

"Calm down Mr Cotton, please," Councillor Dunwell said. He knew his name, as Johnnie had a bit of a reputation for being outspoken and opinionated. He continued: *"Miss Beetledown?"*

Flossie Beetledown continued. *"Thank you for your comments, Mr Cotton. As I said, road safety is of paramount importance and any work we do will have this at its top priority. I also said that only a small percentage of verges need to be cut to make the road safe. I believe the rest should be left natural for the benefit of birds and wildlife and everybody in the long run."*

I had just stopped my performance which I have to say was quite impressive, and was feeling more confident about things now, although it was still not the right time to report back to the others outside.

Councillor Dunwell stood up. He looked purposely around, taking his time to look briefly at everyone. He did this for a whole minute in silence. I got the feeling he was going to say something big and I was ready to react if necessary.

He took a deep breath and said: **"I have listened to you all and thank you for your contributions. Throughout my whole life in public service, I have never encountered anything like this. When I came to Little Snuffling I expected a quiet life and nothing extraordinary to happen. How wrong was I? What I am about to say will probably shock, confound, annoy and hopefully please the majority of members of this audience. I have thought long and hard about this, I've listened to your concerns, complaints, testimonies, emotive outbursts and importantly the surreal and worrying silence of the bird population. I've decided what I am going to do."**

He paused. The room was silent, all eyes trained on him. He reached down for a sip from a glass of water.

"Get on with it!" shouted Johnnie Cotton.

Councillor Dunwell raised his head, and with a loud voice and succinct clarity said: **"I am going to ask the birds what they would like to happen."**

The hall was silent.

The councillor continued: **"This may seem ridiculous, but from what I've heard there may be some truth in their ability to understand we say."**

I was going apoplectic again on my skylight perch, jumping up and down and flapping for all I was worth. This was the opportunity Osbert the Old mentioned! Unfeathered were pointing and smiling at me. Johnnie and Dick weren't, they were making their way to the back of the hall to leave.

NOW was time to report back to the troops! With a well-executed jump and flapping pirouette I was gone. The unfeathered noticed this. A ripple of disquiet and anticipation moved among the audience …

Councillor Dunwell added: *"If these birds can understand what we're saying, let's go and ask them. Let's ask them if Farmer Cobbled and the council plans are the reason they've stopped singing. I think you'll all agree it would be good to get our birdsong back!"* Councillor Dunwell was an astute man who obviously cared about the environment.

I was back with all my bird friends in seconds. *"This is it!"* I said.

"What happened?" said most birds all at once. I only heard Bill, Darcy, Rosie, Jennifer and Osbert the Old as they were nearest but I was assured afterwards that they all spoke as one.

"You were right, Osbert the Old." I said. *"An opportunity has presented itself and we have to act now! Councillor Dunwell believes we can understand them and he is going to ask us what we think of the plans."*

Osbert spoke. *"Well done young Tom, this is more than we could have hoped for. So how do we take advantage of this?"*

I had a plan. We would wait for Councillor Dunwell and the assembled audience to address us. I assumed this would be where we are now as getting all us birds in the hall wouldn't work. *"Right, fellow birds!"* I said *"This is our chance and our future. In a moment Councillor Dunwell is going to come through that door,"* I pointed with my wing to the village hall entrance. *"He is going to ask us whether the Farmer Cobbled's and the council's grass and wildflower verge cutting plans are the reason we have stopped* *singing. When he has finished we all burst into our best song at the tops of our voices. Cain Crow and Rob Rook, you might want to tone the cawing down a bit and leave the singing to the more tuneful brothers and sisters?"*

Cain and Rob looked a bit hurt but understood they had terrible voices and concurred with the plan.

"That should answer Councillor Dunwell's question!" I finished.

We looked at each other with excited faces. We could do this. We could stop our habitat destruction and help our children and future generations of birds. **NOW!**

I decided at that moment that this was to be called **"Tom Thrush ceases the hush!"**, a slight modification on Darcy's clever original name.

Then, Councillor Dunwell led the unfeathered out of the hall. He positioned himself across the road from us and the unfeathered moved alongside and behind him. This was a surreal experience both for us birds and the assembled unfeathered. Arthur Ablebody was wide-eyed and beaming as were his friends Mrs Rightly-Soe and Peter Bumbleton.

I looked across at Farmer Cobbled and his family. He looked calm and ready to accept that his plans may not go ahead. He looked at his wife and kids who were smiling. He himself managed a half smile.

"Are the birds going to sing again, Dad?" asked Ollie.

"I think that's what might happen, Son," said Hugh.

"I can have my cooing woodpigeon back!" said Lucy, with a huge smile.

Flossie Beetledown was gobsmacked by what was about to happen and was taking it all in. Trevor Dredging stood at the back, not quite able to comprehend Councillor Dunwell's decision but at the same time realising he was going to witness something unique.

I was front and centre looking directly at Councillor Dunwell. He had decided I was the leader. Not because of my dancing but mainly because of my presence in the hall and the fact that a couple of members of the audience seemed to recognise me. Councillor Dunwell would address Tom Song-Thrush personally.

He looked at me but said nothing for a while. I think he was trying to process this situation and ensure he knew what he was going to say. I imagined his head was full of many thoughts all at once as he was in a fairly unique position. It was probably the first time an unfeathered had knowingly spoken to a bird and expected them to understand.Talking to birds to ask their opinion on something important was definitely a first for him or anybody!

Councillor Dunwell finally spoke.

"If Farmer Cobbled stopped his plans to remove hedgerows and mow field edges and the council didn't cut the wildflower and grass verges, would you sing again?"

This was it. Time to be heard!

I hadn't sung for nearly four months so I was more than ready. I launched into my loud and clear song with more passion than I had ever before. The others followed as one. **What a sound.**

Breathtaking and eye watering. We sounded like a million symphonies all at once.

The unfeathered took a collective step back in awe.

"I guess that answers my question," said Councillor Dunwell. *"Cancel all farmer Cobbled's plans,"* he said authoritatively, *"and from now on do not cut any wildflower and grass verges other than for safety reasons. Meeting adjourned."*

On hearing this we raised our voices even louder. I looked around. All birds were singing their feathers off with huge smiles on their beaks.

Today was momentous in the lives of birds in and around Elmdown Wood. We had communicated to the unfeathered our concerns and got them to change their environmental decisions to protect us. Bella smiled and looked at me proudly as I stood at the head of all the gathered and singing birds. My kids Tim, Ben, Josie and Jill did the same with a **"That's my dad up there at the front!"**, look on their faces that also said **"Dad is really important today but we're not entirely sure why?"**

I didn't want to make a speech but was thinking how grateful I was for all the support from my fellow songbirds and beyond, particularly Bill, Rosie, and Jennifer, and the wise words from Osbert the Old. I was especially grateful to Darcy whose brilliant idea made this all possible, and finally to my dad.

He had made me aware of our Song-Thrush plight and instilled in me the determination to do my best and try to make a difference. It's a shame he couldn't share with me this special day.

I flew home with my family back to Elmdown Wood, exhausted physically and mentally, and went straight to sleep.

Chapter twenty four
The next generation

I slept soundly that night. It was a peaceful sleep and any dreams I might have had would have been filled with uplifting experiences. I don't remember any however. When I awoke, the first thing I saw was my kids: little Tim, Ben, Josie and Jill sitting on the end of the bed, staring at me with the same **"That's my dad up there at the front!"** look of pride in their faces.

In unison they said, with excited voices, *"Dad, tell us about what happened yesterday, it was obviously very important, please tell us more!"*

I was still half asleep but had the wherewithal to realise that now was the perfect opportunity to share with them some important information. I sat up in bed. I hadn't rehearsed what I was going to say but was realising in the moment that this was going to be a necessary and worthwhile conversation. I tried to choose my words carefully.

"When I was your age or even a bit younger, Granddad Tom used to tell me stories about our family. They always made me a bit sad as they were about how the unfeathered were changing our environment and habitat and affecting our lives. As I grew up Granddad and I talked more about our futures and your futures and agreed to work together to try to stop any further unfeathered interference. Then Granddad was sadly killed."
I stopped here for a moment as my voice started to crack. I think they noticed. Dad hadn't met my kids, which is a regret I have to live with.

"Obviously I was very sad but I made a vow to myself that I would try to stop any habitat change for the benefit of future generations of Song-Thrushes and other birds. What happened yesterday, kids, was the culmination of many months of work. My friends and I successfully stopped some nasty plans by Farmer Cobbled and the council that would seriously affect our food supply and nesting sites."

Little Tim and Ben spoke up together. *"We're proud of you Dad!"* they said with huge smiles on their beaks. *"And we want to help, tell us what to do and we'll do our best."*

"And us," Josie and Jill chorused, *"We are as capable as you boys! We want to help as well."*

I hadn't expected this but it was music to my ears. My kids were proud of me and what I had achieved and were even wanting to help in the future. I couldn't have asked for more. Maybe future generations of Song-Thrushes can build on the relationship we have with the unfeathered?

Maybe we can stop the destruction of our environment for the benefit of all wildlife? Or at least make the unfeathered question why they are doing it?

Who knows? It's a start. I have four determined little Song-Thrushes on board. Let's hope for many more. Now I'm awake, I'm going to fly over to the village, sit high in the old oak and sing for our new found unfeathered friends of Little Snuffling.

I would like to thank them.